The Big Bang
A View from the 21st Century

Springer
London
Berlin
Heidelberg
New York
Hong Kong
Milan
Paris
Tokyo

David M Harland

The Big Bang

A View from the 21st Century

Springer

Published in association with
Praxis Publishing
Chichester, UK

David M Harland
Space Historian
Kelvinbridge
Glasgow
UK

SPRINGER–PRAXIS BOOKS IN ASTRONOMY AND SPACE SCIENCES
SUBJECT *ADVISORY EDITOR*: John Mason M.Sc., B.Sc., Ph.D.

ISBN 1-85233-713-3 Springer-Verlag London Berlin Heidelberg New York Hong Kong
Milan Paris Tokyo

British Library Cataloguing-in-Publication Data
Harland, David M. (David Michael), 1955–
   The big bang: a view from the 21st century. –
   (Springer–Praxis books in astronomy and space sciences)
   1. Big bang theory 2. Expanding universe 3. Astrophysics
   4. Black holes (Astronomy)
   I. Title
   523.1′8

ISBN 1-85233-713-3

A catalogue record for this book is available from the Library of Congress.

Project Copy Editor: Alex Whyte
Cover design: Jim Wilkie
Typesetting: BookEns Ltd, Royston, Herts., UK

Printed in Germany on acid free paper

Edwin P. Hubble

Other books by
David M Harland

The Mir Space Station – a precursor to space colonisation

The Space Shuttle – roles, missions and accomplishments

Exploring the Moon – the Apollo expeditions

Jupiter Odyssey – the story of NASA's Galileo mission

The Earth in Context – a guide to the Solar System

Mission to Saturn – Cassini and the Huygens Probe

with John E Catchpole
Creating the International Space Station

"Let there be light."
God (reputedly)

"To be a cosmologist, you have to know particle physics."
David N. Schramm

"I was very impressed that one simple theory could incorporate so much physics."
Alan H. Guth

"... a phenomenon which is impossible, absolutely impossible, to explain in any classical way."
Richard P. Feynman (speaking of quantum mechanics)

... perhaps *"our universe is simply one of those things that happens from time to time"*.
Edward Tryon (speaking of the Universe as a vacuum fluctuation)

"Galaxies themselves cannot close the Universe."
J.R. Gott

"The most incomprehensible thing about the world is that it is comprehensible."
Albert Einstein

"The more the Universe seems comprehensible, the more it also seems pointless."
Steven Weinberg

"Anyone who isn't shocked by quantum mechanics has not fully understood it."
Neils Bohr

"No one understands quantum mechanics."
Richard P. Feynman

"String theory is a part of twenty-first century physics that fell by chance into the twentieth century."
Edward Witten

"The truth is out there, and we will find it."
Robert Kirshner

"The moment you encounter string theory and realise that almost all of the major developments in physics over the last hundred years emerge – and emerge with such elegance – from such a simple starting point, you realise that this incredibly compelling theory is in a class of its own."
Michael B. Green

"String theory is revealing the deepest understanding of the Universe we have ever had."
Cumrun Vafa

"General Relativity is at least very close to the truth."
Roger Penrose

"I don't pretend to understand the Universe, it's a great deal bigger than I am."
Thomas Carlyle

"I have the result, only I do not yet know how to get <u>to</u> it."
J.K.F. Gauss

"A philosopher once said, 'It is necessary for the very existence of science that the same conditions always produce the same results'. Well, they don't!"
Richard P. Feynman

"The Universe is full of peculiar coincidences."
Martin Rees

"Of all the conceptions of the human mind from unicorns to gargoyles to the hydrogen bomb, perhaps the most fantastic is the black hole."
Kip S. Thorne

"I do not believe that a real understanding of the nature of elementary particles can ever be achieved without a simultaneous deeper understanding of the nature of spacetime itself."
Roger Penrose

"The true delight is in the finding out, rather than in the knowing."
Isaac Asimov

"So-called 'common sense' is definitely detrimental to an understanding of the quantum realm!"
Anonymous

"More of cosmic history occurred in the first millisecond than has occurred in the ensuing 10 billion years."
Edward Harrison

"Science only advances by renouncing its past."
Neils Bohr

"We are insignificant creatures on a small rock orbiting a very average star in the outer suburbs of one of a hundred thousand million galaxies."
S.W. Hawking

"One of the things that makes the history of science so endlessly fascinating is to follow the slow education of our species in the sort of beauty to expect in nature."
Steven Weinberg

"If the Lord Almighty had consulted me before embarking on the Creation, I would have recommended something simpler."
Alfonso of Castile (circa AD 1250)

"Well, the thing about a black hole, its main distinguishing feature, is it's black, and the thing about space, your basic space colour, is it's black. So how're you supposed to see 'em?"
Holly, the computer of JMC's spaceship Red Dwarf

"Space is big. Really big. You just won't believe how vastly, hugely, mind-bogglingly big it is."
The Hitchhiker's Guide to the Galaxy

Contents

List of illustrations

List of tables

Author's preface

The story of the Big Bang encompasses astronomy and high-energy physics. Faced with this broad canvas, an author must select paths that relate how these two communities initially worked independently, and later came to realise that they were approaching the same problem – the origin and evolution of the Universe – from different directions. A century ago, physicists and astronomers thought that only the fine details in their respective disciplines remained to be settled, but each community was shocked to find, in the early part of the twentieth century, that their satisfaction had been premature. Astronomers discovered (a) that the Universe does not consist solely of the stars in the sky and (b) that our Milky Way system is just one in a myriad of galaxies that are racing away from one another in a manner that implied that space itself was expanding. Meanwhile, physicists discovered the shocking nature of the quantum realm. Only towards the end of the twentieth century did they make sense of the forces that determine the characteristics of matter, and they are now seeking the ultimate 'theory of everything'. I will tell how astronomers discovered the vast size of the Universe, that it is expanding, and how the manner in which it is doing so means that it originated in a 'Big Bang'. It has been recently found that, contrary to expectation, the rate of expansion is not slowing but is accelerating. Once again, with their respective 'Standard Models', the particle physicists and astronomers are of the view that they know the big picture and only the fine detail remains to be settled. If they are correct, then good. But are we simply on the verge of another round of frustrating bafflement?

David M Harland
Kelvinbridge, Glasgow
February 2003

Acknowledgements

I would like to acknowledge the help of – in no particular order – Stephen Webb of the Open University in England, Mike Hanlon of the *Daily Mail*, Marc D. Rayman of Caltech, Anthony Fairall of the University of Cape Town, Richard Green of the National Optical Astronomical Observatories, Maarten Schmidt of Caltech, William C. Keel of the University of Alabama, Robert A.E. Fosbury of the European Southern Observatory, Stefanie Komossa and Rainer Schödel of the Max Planck Institut für extraterrestrische Physik in Garching, Germany, John Kormendy of the University of Texas at Austin, Sandra Faber of the Lick Observatory, Wendy L. Freedman of the Observatories of the Carnegie Institution of Washington, in Pasadena, Douglas O. Richstone of the University of Michigan, Saul Perlmutter of the Lawrence Berkeley National Laboratory, Alex R. Blackwell of the University of Hawaii, Robert Gendler, Ken Glover, Bruno Beloff, David Woods and, of course, last but certainly not least, Clive Horwood of Praxis.

Part I

A sense of perspective

1

In the centre of immensities

THE CELESTIAL REALM

It was noticed at an early stage that the sky seemed to comprise 'fixed' and 'wandering' stars, the latter being the planets. A star catalogue was compiled in the third century BC by the Greek philosophers Timocharis and Aristillus. In 134 BC, Hipparchus, the greatest of the classical astronomers, was so amazed by the sudden appearance of a new star which slowly faded and finally disappeared that he drew up a catalogue to assist in the recognition of such apparitions in the future. In addition to noting the positions of some 800 stars, he devised a method of measuring their brightness in terms of 'magnitudes', arbitrarily assigning the terms *first magnitude* to the 20 brightest stars and *sixth magnitude* to the faintest. As everything seemed to revolve around the Earth, it was assumed that the Sun, Moon and planets were riding on transparent crystalline spheres and that the stars were on a 'celestial sphere' set just beyond the furthest planet. Systematic differences between the positions of stars in the earlier catalogue and his own led Hipparchus to conclude that the orientation of the celestial sphere was migrating over a long period, which was disconcerting as the heavens were believed to be 'perfect'. Nevertheless, the stars could reasonably be expected to hold fixed positions with respect to one another, as indeed appeared to be the case. Although the celestial sphere placed all the stars at the same distance, this raised the question of the differences in their brightness. In 1440 Nicholas Krebs, a German archbishop living in Cusa, published a book in which he suggested that the Earth rotated on its axis and moved around the Sun. He also advocated that the stars, which he considered to be similar to the Sun, were randomly distributed through an infinite Universe and only appeared faint because they were far away, with the faintest being the most remote. However, the ancient teachings were so ingrained that he was largely ignored, and when Giordano Bruno at the University of Naples enthusiastically endorsed the idea he was burned at the stake by the Church for heresy. Meanwhile, Nicolaus Copernicus in Poland had suggested that the Moon was the only object to orbit the Earth, and that all the planets, including the Earth, moved around the Sun. This *heliocentric* theory was outlined in a book that was published upon his death in 1543. Although Copernicus dedicated the book to Pope Paul III, the Church dismissed the idea as being a diminution of the status of the Earth in God's Universe. This hypothesis was insightful, but Copernicus retained the idea that the planets moved in perfect circles. In 1609, in analysing the observations of Tycho Brahe, Johannes Kepler concluded that this was not strictly accurate and formulated his results as three 'laws': (1) the planets move in elliptical orbits with the Sun at one focus; (2) a line drawn between a planet and the Sun 'sweeps out' equal areas in equal times, and the planet's velocity is greatest when it is closest to the Sun; and (3) the square of the orbital period of a planet is proportional to the cube of its average distance from the Sun. It was a triumph for astronomers armed with nothing more potent than the naked eye and instruments for measuring angles and the passage of time.

NEWTON'S INSIGHTS

In 1661, at the age of 18, Isaac Newton left his family home in Lincolnshire to study at Trinity College in Cambridge University. His tutor, Isaac Barrow, the Lucasian chair of mathematics, promptly recognised his exceptional mathematical ability, and urged him to investigate the nature of light. Hitherto, this had been debated in a philosophical manner, with little experimentation, and opinion differed over whether it behaved as a liquid or a vapour. Once Newton had read everything that was available on the subject, he set out to devise and conduct his own experiments. Bubonic plague broke out in London in August 1665, and when it showed signs of migrating north, the university sent its students home. As a self-motivated student, Newton spent the next two years in enforced isolation working on his investigations and, if anything, benefited from the absence of distractions. As he broke new ground in understanding not only the nature of light but also how objects moved and interacted with each other, he developed new mathematical formalisms to describe his discoveries.

In particular, Newton noticed that if he darkened his laboratory except for a single ray of light and placed a glass prism into that ray, a coloured band was projected on the wall. While, he was not the first to observe that a prism produced a rainbow, he was the first to realise *why* it did so. At that time, sunlight was believed to be 'pure' white light, with colours being due to 'impurities', but Newton realised that white light did not indicate the absence of colour, it was really a *blend* of colours – ranging from violet through blue, green and yellow to red – which the prism refracted to differing degrees, reacting mostly to violet. He proved this by inserting an inverted prism into the rainbow as it emerged from a regular prism and thereby reformed the ray of white light. He decided to call the artificial rainbow a 'spectrum'.

About 585 BC, Thales of Miletus, the Ionian Greek who was essentially the first scientist, noted that matter appeared to have one of three forms – solid, liquid or gas. In 425 BC, Democritus of Thrace realised that, contrary to popular belief, matter was not continuous but comprised a myriad of discrete and indivisible units, which

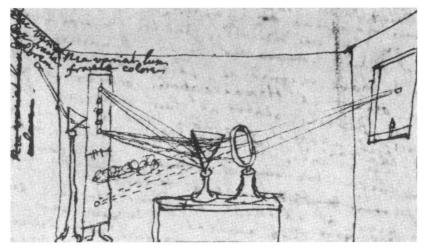

A sketch by Isaac Newton showing how he boarded-up a window to shine a ray of sunlight onto a prism, which refracted the light and projected its spectrum onto a screen. Also shown is the fact that when he placed a second prism behind a hole in the screen it passed only one colour, thereby establishing that the spectrum was not produced by the prism itself.

he named 'atoms' (Greek *atomos*, 'indivisible'). Unfortunately, the great philosopher Plato rejected this in favour of continuous matter, but when Democritus's writings were rediscovered two thousand years later, his idea was accepted. Noting that light did not possess mass, Newton concluded that light comprised tiny massless atoms, or 'corpuscles'. However, he could not explain why a prism refracted corpuscles of different colours to differing degrees, and was puzzled that intersecting rays of light simply passed through one another rather than being diffused as a result of their corpuscles colliding and scattering. Refraction through a lens produced an image with glaring colours, but in 1667 Newton noted that a mirror was free of chromatic aberration, and so he developed a telescope using a curved mirror to focus the light. The Royal Society of London had been established in 1660 to provide a forum for debating and propagating new scientific ideas. After Newton demonstrated one of his telescopes to the Society in 1671, the secretary, Henry Oldenburg, was so impressed that he extended to him an invitation to become a member of the Society. Newton was reluctant, but in 1669 Barrow had relinquished his chair to his pupil and so, as a leading Cambridge don, he accepted and promised to keep the Society appraised of his work. Accordingly, in February 1672 he sent to the Society a paper entitled 'New Theory about Light and Colour', summarising what he had learned from the dozens of experiments he had devised while in isolation from the plague. Unfortunately, Robert Hooke, the curator of the Society – who in 1665 had published an account of his own researches on light – took exception, and wrote a rebuttal. At Oldenburg's behest, Newton engaged in a debate-by-letter which rapidly expanded as more and more people offered their views, and after three years Newton, resenting the time that this consumed, not only ceased to respond but also decided not to communicate his continuing investigations to the Society.

Newton's greatest insight concerned gravitation. The contemporary concept of a force was based on how terrestrial objects moved, and it was believed that physical contact was essential in the application of a force. Gravity had long been recognised as an attractive force, but its application was a mystery since there was no contact with a falling object. In 350 BC, the Greek philosopher Aristotle, a student of Plato, argued that the terrestrial and celestial realms were distinct. As a result, gravity was thought to be a purely terrestrial phenomenon. In 1666, in wondering why an apple fell straight down and yet the Moon revolved around the Earth, Newton realised that a falling apple and the Moon must both feel the *same* force. In formulating his own laws of motion, Newton contradicted the belief that an object moves because a force is applied, concluding instead that an object, once in motion, will *remain* in motion until a force is applied to change this state. On Earth, friction acts to oppose motion, and to maintain an object in motion a force must continually be applied to overcome friction – which explained the earlier misunderstanding. Using the analogy of rocks being thrown at ever greater speeds and falling further and further from the point of origin, Newton realised that the Moon differs from a falling apple by having a horizontal component to its motion. Although it *is* continuously falling, the Earth's spherical surface continuously dips away and the Moon's altitude is maintained. It will therefore revolve endlessly around the Earth precisely because there is no force in the frictionless celestial realm to make it do otherwise. He calculated the strength of the force between any two objects to be proportional to the product of their masses, and inversely proportional to the square of the distance separating them, measured centre to centre. To his delight he discovered that Kepler's empirically derived laws of orbital motion were *necessary*. While it was evident that everyday objects reacted when 'acted upon' by the direct application of a force, it was somewhat more difficult to explain how physically separated objects could influence each other. However, because his equations stiplulated it to be so, Newton introduced the concept of 'action at a distance' and simply refused to speculate about how it operated.

While an undergraduate at Oxford, Edmund Halley wrote a book in which he discussed Kepler's laws of orbital motion. In 1676, John Flamsteed, the Astronomer Royal (in fact, the first of the line), read Halley's book, hired him and immediately dispatched him with a 24-foot telescope to the island of St Helena in the South Atlantic to set up the first observatory in the southern hemisphere. During his two-year stay, and despite poor weather, Halley compiled a catalogue of 341 southern stars. One intriguing object, omega Centauri, was later found to be a huge 'globular' cluster comprised of in excess of 1 million stars. As a result of this dispute with Hooke on the nature of light, Newton did not inform the Royal Society of his gravitational study, and his notes were buried in a rapidly growing stack of paper. When Hooke told Halley in 1684 that he had discovered that gravity obeyed an inverse square law, he was asked to provide a mathematical proof. After much stalling, Hooke had to admit that while he believed this to be so, he could not actually prove it. Shortly thereafter, while paying Newton a visit, Halley enquired whether Newton had given thought to the nature of the force that made the planets orbit the Sun. Newton replied that he had calculated this five years previously.

Although 17,000 lightyears away, omega Centauri (NGC5139) is the brightest of the globular clusters. Visible to the naked eye as a small fuzzy patch, it is 150 lightyears in diameter and hosts in excess of 1 million stars. The stars are packed so close that an astronomer on a planet circling a star in the core would see the sky ablaze with brilliant stars, even during the day. This picture was taken in 1975 by the 4-metre Blanco Telescope at the Cerro Tololo Interamerican Observatory in Chile. Courtesy of National Optical Astronomy Observatories (NOAO), Association of Universities for Research in Astronomy (AURA) and National Science Foundation (NSF).

However, as he was unable to find his notes, he promised to rework the proof and forward it. Halley was so impressed with the elegance of the method that he urged Newton to write up in book form *everything* that he had discovered of the laws of motion. Newton agreed. In 1670 Newton had invented a mathematical formalism that he called 'fluxions' (known to us as 'calculus'). He had used that method in his investigation of motion, but for the book he laboriously proved all his theorems in traditional geometric style in the hope that this would help others to follow his logic. As it was such a scholarly tome, he wrote the work in Latin. When *Philosophiae Naturalis Principia Mathematica* appeared in 1687, Hooke claimed to have been the first to prove the inverse square law. As a result of this second incident, Newton did not publish *Optiks*, the comprehensive record of his investigation of the nature of light, until 1704, a year after Hooke's death. After using Newton's law of gravitation to calculate the orbits of historical comets, Halley suspected in 1705 that several viewings with essentially the same highly elliptical orbit were repeated appearances of a single comet. He predicted that it would reappear in 1758 – which it did – and although he died in 1742 and did not witness the return of the comet, it now bears his name: Halley's Comet. This triumphant prediction banished any lingering doubt as to the validity of Newton's law of gravitation.

THE SOLAR SYSTEM

Copernicus had been able to make a reasonable estimate of the *relative* sizes of the orbits of the planets around the Sun, and Kepler refined his figures. If any specific distance could be measured, this would provide the scale of the Solar System. Hipparchus had trigonometrically measured the distance to the Moon and, following an eclipse, had calculated that the Sun was 37 times further away – at the then almost unimaginable distance of 15 million kilometres – but this was only a crude estimate.

C.D. Cassini, the Italian director of the Paris Observatory, decided to determine the distance to Mars during that planet's favourable opposition in 1672. He sent his assistant, Jean Richer, to French Guiana in South America with instructions to make a series of measurements of the planet's position in the sky. Trigonometry over this 10,000-kilometre baseline gave an absolute measure with which to calibrate Kepler's relative scale. The result showed the Earth to orbit the Sun at a distance of about 136 million kilometres, and this yardstick – the Astronomical Unit (AU) – was adopted for measuring celestial distances. On this scale, the planet Saturn, which was at that time the most distant member of the Solar System, was an incredible 1 *billion* kilometres distant. And what of the stars beyond? Christiaan Huygens, the leading Dutch astronomer, argued that if Sirius, the brightest star in the sky, was as luminous as the Sun, then to appear as faint as it does it must be about 30,000 times as distant, but it was difficult to measure the relative brightnesses of the Sun during the day and Sirius at night. Could the distance to a star be directly measured? Although measuring the parallax of a star was straightforward in principle, it would require the longest possible baseline, which was the diameter of the Earth's orbit, with observations made six months apart. Copernicus had noted that the *absence* of

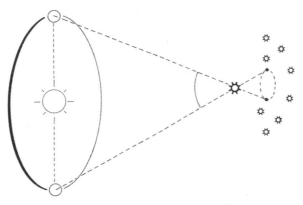

As the Earth pursues its orbit of the Sun, a nearby star will appear to oscillate against the background of more distant stars. The distance to that star can be calculated trigonometrically.

parallax meant that the stars must be *very far* away. The precision required was beyond the naked-eye observer with a quadrant, but the invention of the telescope in the early seventeenth century offered tremendous prospects.

In 1710 Halley took an interest in the classical star catalogues. It had long been known that the positions of the stars migrate in a systematic manner. Halley realised that this implied that the Earth's axis of rotation was 'precessing' with a period of about 25,000 years. The north celestial pole is currently near a star appropriately named Polaris. In 4600 BC, when the Egyptians were building the pyramids, the pole was near Thuban in Draco. Twelve thousand years before that, it was near the bright star Vega in Lyra. In 1718, in comparing his observations with Hipparchus's catalogue in order to refine the precessional rate, Halley noted that the bright stars Sirius, Aldebaran and Arcturus all showed 'proper motions'. Sirius was travelling across the sky at 1.3 seconds of arc per year. Since classical times, it had travelled a distance equivalent to one and a half times the angular diameter of the Moon. In fact, Sirius's progress was so pronounced that it was measurable since Brahe's observations 150 years earlier. The discovery that stars have *independent* motions through space marked the downfall of the concept of the celestial sphere. The fact that some stars moved more than others hinted that they were situated at different distances, as Krebs had speculated three centuries previously. And the fact that the *brightest* stars showed such motions suggested that they were closest to the Earth. Having tried and failed to discern the parallax of Sirius, Halley reasoned that if it was as luminous as the Sun then for it to appear as faint as it does it must be 120,000 times as far away. At this distance its parallax would be 1 second of arc, some 200 times finer than had been attained by Brahe using his large quadrant, and too small an angle for Halley to measure using his telescope.

James Bradley set out in 1725 to measure the parallax of gamma Draconis. He chose this star because it was bright and because it passed almost directly over London every night of the year. Bradley was therefore able to fix a 24-foot telescope vertically on the chimney of the house of his friend Samuel Molyneux in order to eliminate any effects that might influence the optical configuration on successive nights. This had been attempted by Robert Hooke a century previously, but with inconclusive results. Over a year, Bradley did indeed observe changes in the star's position, but it was *not* parallax. In 1675, while pondering why the tables that he had drawn up to predict the times when Jupiter's satellites should enter and exit its shadow were flawed, the Danish astronomer Olaus Römer, working in Paris, had realised that his tables were correct and, contrary to contemporary wisdom, light travelled at a *finite* speed. From the varying distances between the Earth and Jupiter as they independently orbited the Sun, Römer calculated that light travels at approximately 300,000 kilometres per second. Bradley realised in 1728 that the effect he saw was a consequence of light's finite speed. The effect is most readily understood by an analogy: rain that is seen to fall vertically by a stationary observer will appear to a moving observer to be falling at an angle. This 'aberration of light' (as Bradley called it) *proved* to any remaining sceptics that the Earth really was travelling through space. As this motion inclined the path of a light ray travelling down the tube of his telescope, when he centred a star in the field he was neither really pointing at it nor measuring its true position in the sky. The displacement could be as much as 20 seconds of arc. After accumulating 20 years' of observations, Bradley found another complication in the form of 'nutation' in the Earth's axis of rotation – a slight wobble that, he reasoned, was the result of the world not being a perfect sphere. After eliminating these effects, he was disappointed to find that there was no residual evidence of a parallax. Believing that he could measure a position to 1 second of arc, he concluded that the absence of a parallax meant that this star had to be *very far* away. This proved that the stars were at least comparable to the Sun in luminosity.

A SYSTEM OF STARS

Democritus of Thrace speculated in 425 BC that the Milky Way was a system of stars, but this idea was not popular. When Galileo Galilei aimed his first telescope at the Milky Way in 1609, he was astonished to resolve the faint glow into a myriad of stars. In *An Original Theory or New Hypothesis of the Universe*, published in 1750, the English surveyor Thomas Wright argued that the Sun did not lie at the centre of the Universe, but was one of the stars on a sphere so immense that we perceive only the local part of it as the Milky Way. In 1751 Immanuel Kant, who would later become a renowned philosopher, encountered a summary of Wright's book that focused on the final aspect of the hypothesis, and gave the impression that the Milky Way was a disk-shaped system which extended to infinity. Kant found this appealing, and in 1755 he wrote *Universal Natural History and Theory of the Heavens*. In 1742 the Frenchman Pierre Louis de Maupertius had written of having seen small luminous patches in the sky, the most prominent example of which, in Andromeda, could be glimpsed by the naked eye on a dark night. Kant speculated that these were independent systems of stars, similar in nature to the Milky Way, which he dubbed "island universes". Unfortunately for Kant, his publisher was bankrupted before the volume appeared. In 1764 J.H. Lambert told the Berlin Academy of Sciences with great conviction that the Milky Way was a disk-shaped system of stars, that the Sun was located towards its edge, and that there were innumerable other such systems in the furthest reaches of infinite space, but his exposition fell on deaf ears. After independently spotting the comet that Halley had predicted would return in 1758, Charles Messier in France started hunting for comets. In 1771 he circulated to fellow comet-seekers a list of 45 fuzzy objects that were all too readily mistaken for remote comets. A decade later the French Academy of Sciences published a revised list with just over 100 such 'nebulae'.

After several frustrating years of making and using long and unwieldy refractors, William Herschel switched to reflectors and in 1774 made one with a mirror that was 6 inches in diameter and had a focal length of 7 feet. This reflector proved to be superior to the telescope used by the Astronomer Royal. On 13 March 1781, while charting the sky, Herschel noticed a small sea-green disk. Presuming it to be a tail-less comet, he noted its motion against the stars for several nights before reporting his discovery, which was soon realised to be an unsuspected planet. Herschel named it *Georgium Sidus* in honour of King George III, but continental astronomers objected and, following the suggestion of J.E. Bode of the Berlin Observatory, the planet was later named Uranus. In the autumn of 1782 Herschel received a copy of Messier's list of nebulae from Alexander Aubert, an astronomer friend. He promptly set about studying these objects using his 'small 20-footer' (a reflector with a 12-inch mirror) and found that while many of them had been listed by Messier as diffuse, he was able to resolve them as clusters of stars. In October 1783, when he finished his 'large 20-footer' (with its excellent 19-inch mirror) he was able to resolve even more. Speculating that they were so far away that they were *independent* systems, "some of which may well exceed our Milky Way in grandeur", he postulated that our system would appear as a fuzzy patch in the telescope of an astronomer on a planet orbiting a star in such a cluster. So motivated, he set himself the momentous task of "gauging the construction of the heavens". When explorers ventured into far southern latitudes, they discovered that the Milky Way girdled the sky. Herschel decided to discern its shape in terms of the distribution of stars, but as it was clearly impractical to count *all* the stars he chose 683 sample areas, and over an 18-month interval in 1784–1785 he counted the numbers of stars present of each magnitude. In his analysis, he reasonably assumed that the stars were evenly distributed in space. "The Milky Way is undoubtedly a most extensive disk of stars", he wrote in 1785, and judged the Sun to be located near the disk's centre. Unlike Wright, Herschel did not presume the disk to extend off to infinity. On his methodical "sweeps" of the heavens, he saw many celestial objects whose characteristics led him to suspect an evolutionary sequence in which stars congregated in loose clusters that shrank into progressively more compact forms. In a radical departure from classical thinking, he mused whether the Milky Way was *evolving* similarly, and he thereby presented the first model of its origin and fate. Herschel estimated the disk's diameter to be 850 times the distance to Sirius, and one quarter of this size at its thickest point. To determine the scale of the Milky Way system, Herschel decided to try to measure the distance to a star.

As Astronomer Royal, Flamsteed had started to compile a star catalogue, but he died before all the observations could be processed. This task was completed by Halley and the catalogue was published in 1725. Although it was an excellent piece of work for the early eighteenth century, its star positions were accurate only to 10 seconds of arc, which was inadequate for measuring a parallax of at most one-tenth of that angle. Galileo had reasoned that if all stars were of the same luminosity, their relative brightnesses were due to their differing distances. He also reasoned that as the Earth moved around the Sun, the brighter stars ought to be displaced more than the fainter ones, and hypothesised that a series of measurements of the positions of a bright star relative to a fainter one on almost the same line of sight should permit the distance of the brighter star to be determined. This had been beyond Galileo's technology, but Herschel was confident that by fitting a micrometer to the eyepiece of his telescope he could accurately measure the angular separation between two stars in the same field of view. In 1784 he began to monitor the relative positions of some 400 'double' stars. To his delight he detected motion, but by 1793 it was evident that this was not parallax. By 1804 the motions of three cases were sufficiently extensive to indicate that the pairs of stars were physically related, in *binary* systems. This finding revealed that all stars are *not* of the same luminosity, and proved that Newton's law of gravity was truly universal. Herschel continued observing in the hope of identifying parallax, but in vain. In addition to extending the catalogue of proper motions, in 1805 he decided to search for a systematic drift that could be attributed to the *Sun's* motion through space. The fact that the Earth's orbital motion seemed to produce *no* measurable parallaxes simplified his task. The 13 proper motions measured by Nevil Maskelyne and J.J. Lalande gave an indication that the Sun was moving in the general direction of the constellation of Hercules. When Maskelyne issued an extended list of 36 proper motions, Herschel repeated his analysis and refined the 'apex' of the Sun's motion as lying on the border between Hercules and Lyra. Furthermore, this revealed that contrary to expectation the stars were not moving in random directions but, to some extent, *shared* the Sun's motion, suggesting that the Milky Way system was in a state of rotation. After the previous discovery that some of the nebulous patches on Messier's list were clusters of stars, Herschel presumed that a larger telescope would establish this to be the rule. Although a grant from the Crown enabled him to build a telescope in 1789 that had a mirror 48 inches in diameter, a focal length of 40 feet, and an increased 'light grasp' that made the nebulae brighter, many remained unresolved. In undertaking his sweeps, Herschel catalogued many star clusters and fuzzy patches. He sent a list of 1,000 to the Royal Society in 1786, 1,000 more in 1789, and a further 500 in 1802. Many of the objects that Herschel resolved into systems of stars are now known as 'globular' clusters.

THE LIGHTYEAR

While cataloguing proper motions in 1792, Giuseppe Piazzi of Palermo, Sicily, discovered that the fifth-magnitude star 61 Cygni was travelling at 5.2 seconds of arc per year, giving rise to the nickname the 'flying star'. Having fled Germany after Napoleon's invasion in 1808, F.G.W. Struve entered the University of Dorpat (now Tartu in Estonia). A decade later, he was appointed director of its observatory. The 9.5-inch Fraunhofer achromatic refractor had the largest lens in the world. Its 'equatorial' mount, the first of its kind, enabled it to be readily directed towards any point in the sky, and a clockwork drive compensated for the Earth's rotation. It was ideal for making fine measurements of the positions of double stars. In 1830 Struve began to monitor 61 Cygni – a double separated by about 27 seconds of arc – and his results indicated that they revolved around a common centre in approximately 650 years.

In 1804 F.W. Bessel, then aged 20, wrote a paper in which he recalculated the orbit of Halley's comet. He dispatched a copy to the renowned astronomer Heinrich Olbers, who arranged its publication and provided an introduction to J.H. Schröter, a magistrate and astronomer in Lilienthal, near Bremen. Schröter hired Bessel as an assistant at his own observatory, and when the Prussian government founded the Königsberg Observatory in 1810 Bessel was appointed director. In 1818 he published a catalogue of 50,000 stars noting, where possible, their proper motions. In 1829 he installed a Fraunhofer refractor that was so good that he decided to try to measure the parallax of a star. He selected 61 Cygni on the assumption that, despite being faint, its high proper motion meant that it was nearby. He began observations in 1834, but was distracted by the return of Halley's comet the following year and did not resume his parallax study until August 1837. Determining this star's parallax was complicated by the fact that it was moving so rapidly. His task was to isolate the tiny wobble in its track across the sky. F.W.A. Argelander at the Bonn Observatory had worked as Bessel's assistant and gained his doctorate from Königsberg in 1822. Bessel requested that he search through the archives to trace 61 Cygni's proper motion back to 1755. Within a few months, Bessel was sure that there was indeed a wobble in its path, but at least six months of observations were needed to measure a parallax angle. The result was 0.31 second of arc, placing the star at a distance of some 660,000 AU. In order to check for unrecognised systematic effects Bessel completed the year and, when satisfied, announced his success in December 1838.

In 1831 Thomas Henderson, an amateur astronomer who earned his living as a legal clerk in Dundee, Scotland, was appointed director of the newly established observatory at the Cape of Good Hope in South Africa – the first major observatory in the far southern hemisphere. Henderson's first task was to update the catalogue compiled by Halley on St Helena 150 years earlier. On finding that alpha Centauri had a proper motion of 3.7 seconds of arc per year, he concluded that it must be close – an inference that was supported by the fact that it was the third brightest star in the sky. Furthermore, alpha Centauri was a multiple system in which the motions of its components were readily discernible. This meant that they were physically close to one another. Their wide separation in the sky was therefore another indication of the system's proximity. In April 1832 Henderson began a series of observations to try to measure the parallax of the main component; however, he found life at the Cape depressing (deriding the observatory's location as "a swamp") and in 1834 accepted the post of Astronomer Royal for Scotland. Once established in Edinburgh, he analysed his observations and calculated a parallax of 0.9 second of arc. As a check, he requested supplementary observations from a former colleague at the Cape and published his result on 9 January 1839, a month too late to claim priority for being the first to achieve such a measurement.

Between 1835 and 1838, F.G.W. Struve had tried in vain to measure the parallax of Vega, which he had selected because it was the fourth brightest star in the sky and had a proper motion of 0.35 second of arc per year. When Czar Nicholas I built the Pulkova Observatory south of St. Petersburg in 1839, he made Struve its director. After monitoring Vega for a year through his new 15-inch Merz refractor, Struve announced in 1840 that its parallax was 0.29 second of arc, placing it only slightly further away than 61 Cygni. As the Earth–Sun distance was too small a unit for the stellar realm, it was decided to use the 'lightyear', i.e. the distance travelled by light in one year. Because the speed of light is almost 300,000 kilometres per second, this gave the lightyear a value of 10 *trillion* kilometres.

Although Bessel's parallax for 61 Cygni stood the test of time, and the star is almost 11 lightyears distant, a subsequent study halved Vega's parallax, increasing its distance to 27 lightyears. Henderson's parallax was also reduced, to 0.7 second of arc, placing alpha Centauri at a distance of 4.3 lightyears. By the mid-nineteenth century, it was evident that unless there were faint stars with even greater proper motions, the Solar System – which a century earlier had seemed to be vast – was actually *minuscule* in comparison to the stellar realm. Once Sirius was calculated to be 9 lightyears distant, it became possible to calibrate Herschel's scale: at 7,500 lightyears across, the disk of the Milky Way system was *enormous*.

Part II

The forces of nature

2

The mysterious aether

THE NATURE OF LIGHT

Born in 1629 into a notable Dutch family with a tradition of diplomatic service to the House of Orange, Christiaan Huygens studied law and mathematics, then his father provided a stipend to enable him to devote himself to the study of nature. In 1678, in studying optics in ignorance of Newton's corpuscular theory, he inferred light to be a 'wave' in an all-pervasive but otherwise imperceptible medium. In the belief that the celestial realm was 'perfect', and abhoring the concept of a vacuum, Aristotle had introduced the word 'aether' for the transparent frictionless substance that he believed must pervade the celestial realm. It seemed, therefore, that light was a vibration in the aether. If this were proved to be true, then the aether could not be confined to the celestial realm, and must be all pervasive as a substrate to matter. A longitudinal wave, such as sound, can transit solid, liquid and gaseous materials, but transverse waves can pass only through solids. The aether seemed to be an extremely rarefied gas, with light a longitudinal wave. Huygens also reasoned that if light travelled more slowly in a dense medium (such as glass) than in air, then a prism would refract light of different 'wavelengths' to different degrees – the shorter wavelengths being refracted most. This implied, in turn, that the wavelength of violet light must be less than that of red light. However, Huygens was puzzled by the fact that while the waves from an object dropped into water washed around any obstacles and encroached upon the 'shadow' beyond, such diffraction did not render fuzzy edges to shadows cast by the Sun.

The corpuscular and wave theories could each explain some observations but not others, and the phenomena that they explained tended to be mutually exclusive. The debate raged during the eighteenth century, with opinion tending to follow national allegiances but, largely due to Newton's stature, the corpuscular theory came to the fore. However, in 1801 Thomas Young in England discovered that a pattern of alternating light and dark bands was made when light was passed through a pair of narrow slits spaced slightly apart. Such 'interference' was inexplicable if light was

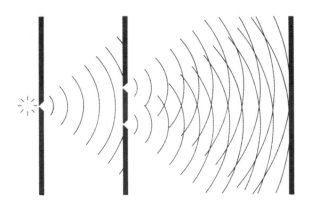

In 1801 Thomas Young discovered that light passing through a pair of narrow slits produced an 'interference' effect which was inexplicable if light was corpuscular but readily explained if light was a wave.

corpuscular, but it was readily explained if light was a wave. By analysing the interference patterns, Young calculated the wavelength of white light to be of the order of 5×10^{-7} metre. This explained Huygens's problem with sharp shadows, because the diffraction of light is negligible for an everyday object.

In 1669 Erasmuth Bartholinus in Holland had noted that light passing through a crystal of calcium carbonate was split into two rays, which it refracted to differing degrees. Neither Newton nor Huygens had been able to explain this, but in 1808 É.L. Malus in France discovered that when a ray of sunlight was passed through a doubly refracting crystal after reflecting off a mirror, the crystal reverted to 'normal' and produced only one refraction. From this, Malus supposed that light possessed two 'polarities'; that whereas normal crystals did not split them doubly refracting ones did; and that mirrors reflected items selectively. Such 'polarisation' could not be explained by the corpuscular theory. A mathematical study of *transverse* waves by A.J. Fresnel in France in 1814 established that sunlight is unpolarised and that the act of reflection polarises light, exactly as occurs in the crystalline structure of a doubly refracting crystal. This raised a serious problem, because the rate at which a transverse wave propagates is dependent upon the rigidity of the medium, and for the wave to travel at the measured speed of light the aether would have to be as rigid as refined iron. It was therefore supposed that there were several components to the aether, and that the 'luminiferous aether' that carried light was extremely rigid. Thus the mid-nineteenth century physicists were faced with the strange concept of a 'rigid gas'.

About 1850, the Swedish astronomer A.J. Ångström suggested that wavelengths should measured in units of 10^{-10} metre, with the range extending from 3,800 Å for violet to 7,800 Å for red. As there was no reason to believe that these were fundamental limits, it seemed plausible that there should be longer and shorter waves to which the human eye was not sensitive. While experimenting with filters to assist his observations of the Sun in 1800, William Herschel had discovered that some materials passed light but not radiant heat, while others, which seemed to be opaque, passed radiant heat. When he projected a solar spectrum and measured the temperature along its length, he found that the peak in temperature was *beyond* the red. The radiant heat was being refracted by the prism, suggesting that it was an 'infrared' wave. In 1801, J.W. Ritter in Germany discovered 'ultraviolet' waves in sunlight, showing again that the spectrum was much broader than the visible range.

ELECTRICITY AND MAGNETISM

About 600 BC, Thales of Miletus conducted a study of the manner in which some objects attract others, and discovered that amber resin (for which the Greek word is *electra*) had the unusual property that, when rubbed, it temporarily acquired the ability to attract paper and other lightweight objects. He was also interested in an iron ore known as 'lodestone', which strongly attracted objects made of refined iron. Since his sample came from the nearby town of Magnesia on the Aegean coast of what is now Turkey, he called it 'Magnesian rock', which later gave rise to the name 'magnet' and the phenomenon of 'magnetism'. Not unreasonably, he considered these phenomena to be distinct.

Peter Peregrinus in France discovered in 1269 that a magnet had 'ends', and that pairs of magnets would attract one another only in a specific alignment. If one of the

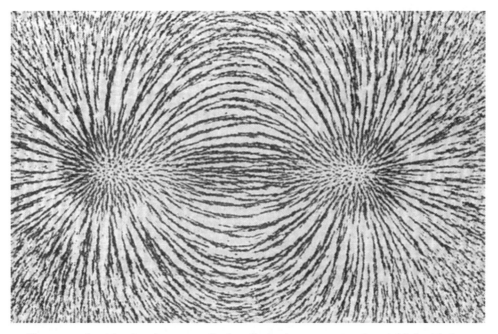

If a paper sheet is placed on top of a lengthwise bar magnet and sprinkled with iron filings, these assume a striking pattern of arcs.

magnets was rotated to present its opposite end, the magnets would repel each other. He also discovered that when iron filings were scattered on a piece of paper that was positioned over a magnet they took on a striking pattern of arcs. It was also observed that a magnetic needle that was free to rotate on a piece of cork afloat in a tank of water would always point towards the north. On his voyage westward across the Atlantic in 1492, Christopher Columbus saw that the deviation of his 'magnetic compass' from true north (determined using the stars) varied with location. William Gilbert, the physician to the court of Queen Elizabeth, had an amateur interest in physics. In 1600 he discovered that a freely mounted compass that could rotate vertically as well as horizontally also 'dipped', from which he concluded that the Earth itself was a giant magnet. This explained the origin of the lodestone rocks. He also reasoned that substances which displayed an attractive force when rubbed became 'electrified' by acquiring an 'electric charge', and since these charges were 'static' (by which he meant that they were stationary) the phenomenon was named 'electrostatics'. However, Stephen Gray in England suggested in 1729 that an electrified fluid was involved, and flowed as an object was 'discharged'. After exhaustive tests he found that there were in fact 'conductors' and 'insulators'. In 1733 C.F. Du Fay in France realised that there were two types of electric charge. A decade later Benjamin Franklin, in America, realised that in combination these charges cancelled each other, and introduced the terms 'positive' and 'negative'. The first serious attempt to *measure* the strength of the electrostatic force was made in 1785 by C.A. Coulomb in France, who discovered that it varied inversely with the square of the separation. The same relationship applied to magnetism. As gravitation also followed this law, it appeared to be a fundamental property of forces.

A 'condenser' to store an electric charge was invented in Germany in 1745 by E.G. von Kleist, but when he received a tremendous shock while discharging it, the device was discarded. A similar device was independently developed the following year by Pieter van Musschenbroek at the University of Leiden in Holland. Although he also suffered a shock, he persevered and perfected the 'Leiden jar', which facilitated much of the subsequent research into electricity. After reflecting upon the similarity of the discharge of a Leiden jar to a bolt of lightning, Franklin flew a kite during a thunderstorm in 1752 in the hope of drawing lightning to charge his condenser – which he did with near-fatal consequences. Discharges were brief violent currents. In 1800 Alessandro Volta in Italy was the first to make a charge flow in a sustained 'current' through a wire. Upon hearing of Volta's achievement, William Nicholson and Anthony Carlisle in England discovered that passing a current through water dissociated its molecules in a process called 'electrolysis'. In 1812 Michael Faraday was hired as an assistant in the laboratory of the Royal Institution by the chemist Humphrey Davy, and by 1825 he was the laboratory's director. After a thorough study, Faraday realised in 1832 that the mass of water electrolysed was proportional to the total amount of electricity used, and this discovery laid the basis for electrochemistry. Realising that there was a *unit* of electric charge, he introduced the term 'ion' (which is Greek for 'wanderer') for the electrically charged entities that electrolysis produced – although he had no idea what they were. H.C. Oersted in

Copenhagen serendipitously observed in 1819 that an electric current in a wire induced a deflection in a magnetic needle, indicating that electricity created a magnetic field and showing that there was a relationship between the two forces. In France, D.F.J. Arago used iron filings to show that the magnetic force from a current-carrying wire was indistinguishable from that of a magnet. In effect, the wire *became* a magnet when a current flowed. In 1820 A.M. Ampère found that current-carrying wires attracted and repelled each other in the manner of magnets, and Faraday promptly exploited this discovery to create an electric motor. After a study of how magnets with shapes ranging from linear bars to horseshoes distributed iron filings, Faraday realised that there was a regularity to the patterns. By moving a compass needle around the magnets, he also discovered that the patterns were three dimensional. In his book *Theoria Philosophiae Naturalis* published in 1758, the great Dalmatian physicist R.G. Boscovich had ventured that all objects are surrounded by 'force fields' that could influence nearby objects. In 1831 Faraday reasoned that the 'lines of force' depicted in iron filings were a physical manifestation of the magnetic force, and this offered a *mechanism* for Newton's concept of 'action at a distance'. If a magnet emitted a field that pervaded the surrounding space, then another magnet located within the field would react to the field's presence. Even though they were some distance apart, the magnets would 'act' upon each other. Since magnetism generated an electric current and a current generated magnetism, Faraday devised an experiment to determine whether magnetism could *transmit* electricity. He wound a wire around one half of an iron ring and another wire around the other half, and found that the act of switching the current on and off in one wire induced a momentary current in the other. As there was no such 'induction' when the current was flowing smoothly, he realised that it was the onset or collapse of the magnetic field that caused the effect; the current in the second wire was induced by the *change* in the magnetic field. Later, reversing the principle of the motor, Faraday developed a device in which a rotating magnet induced a sustained electric current – the electric generator.

Otto von Guericke in Germany had invented a device in 1650 to extract air from a container. As the apparatus was improved, he was able to create ever more rarefied environments. In 1783 French scientists began to send instruments aloft by slinging them under balloons of hydrogen gas, and discovered that the pressure decreased markedly with altitude. The revelation that the atmosphere is no more than a thin aura to the Earth meant that there must be a vacuum beyond, which meant that light and gravitation must be able to propagate through a vacuum. Experiments using evacuated containers showed that electric and magnetic forces were still present. However, since the mid-nineteenth century physicists insisted that these forces *must* propagate through a medium, they chose to define 'vacuum' as the absence of matter; the aether, they argued, was the essence of the vacuum. The aether simply *had* to exist, for otherwise the starkness of the vacuum would reintroduce the unsatisfactory concept of 'action at a distance'. Faraday's lines of force must therefore be distortions in the aether.

ELECTROMAGNETISM

James Clerk Maxwell developed an early aptitude for mathematics and published his first scientific paper in 1845 at the age of 14. Two years later he was accepted by Edinburgh University to study physics, and in 1854 he attended Trinity College, Cambridge. Lacking mathematical training, Faraday had drawn upon an analogy with a rubber band to explain his 'lines of force' in terms of distortions of the aether. In 1855 Maxwell set out to develop a mathematical description of Faraday's insights, modelling the force as streams of a fluid that flowed around and between objects. In 1862 he refined this analogy to explain both electricity and magnetism in terms of swirling vortices in this hypothetical fluid. By 1864 he had developed four equations that not only described *all* known properties of electricity and magnetism – including the way in which an electric current generated a magnetic field and how a magnet induced an electric current – but also established that electric and magnetic fields could not be considered in isolation; in fact, they were both aspects of the single phenomenon of 'electromagnetism'. As had been discovered experimentally, it was the *change* in the strength of a magnetic field that induced an electric current – and vice versa. This was reflected directly in the electromagnetic field in which transverse waves oscillated in planes orthogonal to the direction of propagation, giving rise to a self-sustaining waveform whose wavelength was defined by the rate at which the component fields waxed and waned.

When Maxwell calculated the speed of an electromagnetic wave, he found that it was precisely the same as the speed of light. Rejecting the possibility that this was coincidental, he inferred that light was an electromagnetic wave, thereby banishing the concept of 'action at a distance' from electricity, magnetism and light. The idea that light was a manifestation of electromagnetism was so astonishing that few people took it seriously. The equations indicated that when an electric charge was accelerated, an electromagnetic wave would propagate outward in an ever-expanding spherical shell. Since the wave *radiated* away from its source, this phenomenon was called 'electromagnetic radiation', and although this implied that electromagnetic waves would propagate energy through space, Maxwell never felt

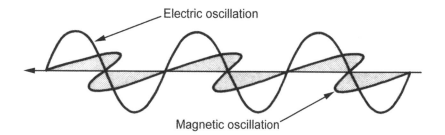

Electromagnetism is a self-sustaining waveform in which the associated electric and magnetic fields oscillate in a transverse manner out of phase in planes orthogonal to the direction of propagation.

the need test this experimentally. However, in 1888 H.R. Hertz of the Karlsruhe Technical Institute in Berlin generated an electrical spark in an appartus on one side of his laboratory and noticed a weak signal in a detector on the other side of the room, thereby confirming that the spark had propagated electromagnetic energy. Although this meant that light was indeed an electromagnetic wave, the fact that he had not seen the wave cross the darkened room implied that, in this case, it was at a wavelength that did not fall within the visible range. Hertz found that the strength of the induced signal varied in a periodic manner with the distance between the source and his detector, and was thus able to measure its wavelength. These long-length waves were initially referred to as 'Hertzian waves', but after Guglielmo Marconi demonstrated 'radio telegraphy' in 1896, they were renamed 'radio waves'. Unfortunately, Maxwell did not live to see his theory so spectacularly verified.

RELATIVITY

A.A. Michelson reasoned that if electromagnetic waves were oscillations in the luminiferous aether, it ought to be possible to measure the Earth's motion through it. In 1881 he erected a split-beam interferometer in Helmholtz's laboratory to measure the speed of light on orthogonal axes. Upon failing to detect 'aether drift', he decided that the instrument was insufficiently precise. In 1886 he was a professor at the Case School for Applied Science in Cleveland, Ohio, and together with E.W. Morley he re-engineered the instrument and emplaced it in bedrock for stability. Although the instrument functioned satisfactorily, they were obliged to announce in 1887 that there was *no* measurable drift. In 1893 the Irish physicist G.F. FitzGerald suggested that because the aether pervaded all matter, the Earth's passage through it induced a pressure which distorted objects, foreshortening them along the direction of motion and, as a consequence, *measurements* of the speed of light along the direction of the Earth's motion *should* be the same as the speed measured perpendicular to it. In his view, the 'unexpected' result of the Michelson–Morley experiment was proof that the aether existed! After the Dutch physicist H.A. Lorentz refined the mathematics, this effect became known as the Lorentz–FitzGerald contraction. Lorentz found that the mass of an object would appear to increase in proportion to its motion through the aether. FitzGerald also noted that the mathematics implied that the speed of light was a fundamental limit – and that nothing can be accelerated through this speed: in effect, as the mass increased, the force needed to accelerate it increased until, just short of the speed of light, an infinite force would be needed to attain and exceed this speed.

In 1900 Henri Poincaré in France dismissed the aether as an 'abstract' frame of reference without physical significance. Two years later, he declared the issue of the existence of the aether to be an exercise in metaphysics of no importance. In 1904 he said that it was impossible to detect *absolute* uniform motion, and hence, contrary to Newton's presumption, there was no unique frame of reference; all frames are valid, and each observer is free to choose one that is convenient. Poincaré expressed this as the Principle of Relativity, which stated that the laws of physics were independent of the frame of reference used by an observer.

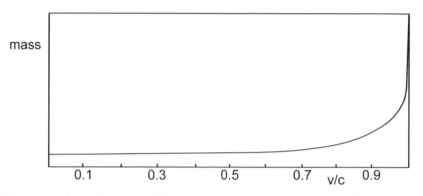

The mass of an object travelling relative to the local frame of reference appears to increase, but the effect does not become significant until the speed approaches the speed of light. This is known as the 'Lorentz factor'.

Having unified electricity, magnetism and light, Maxwell reflected that although electromagnetism is inherently stronger than gravitation, both varied similarly with distance, and although he speculated that there might be a fundamental relationship he did not pursue the matter. After scraping through college and failing to secure a teaching position, Albert Einstein became, in 1901, an examiner at the Patent Office in Berne, Switzerland, where, in his spare time, he conducted 'thought experiments' exploring how Maxwell's theory might be enhanced so as to embrace Newton's gravitation. In light of the Michelson–Morley experiment, which he interpreted to mean that the aether did not provide a frame of reference in which absolute motion could be measured, he independently developed Poincaré's Principle of Relativity. In a paper published in 1905, Einstein dismissed the aether, and showed how the assumption that the speed of light was independent of the frame of reference (that is, measuring it always gave the same value) led inexorably to the conclusion that a rapidly moving object would appear to become foreshortened and gain mass. Furthermore, he found that the pace of time also appeared to slow down for a rapidly moving object. Of course, from the point of view of that object, using its own frame of reference, this impression would be reversed. Because he considered only the special case of how the world would appear to observers in two frames of reference that were moving with respect to one another at a *fixed velocity*, this concept was referred to as the Special Theory of Relativity. As part of this work, he also showed that energy and mass are equivalent by the equation $E = mc^2$, and introduced the unifying concept of 'mass-energy'; mass being a concentrated form of energy, because the speed of light, c, is a large number and c^2 is a *very* large number. A rapidly moving object appears to increase its mass because its kinetic energy must be included in its overall mass-energy. Insightful though special relativity was, it did not address gravitation, so in 1907 Einstein began to investigate how the theory could be extended. Musing upon the observation that a man riding in an elevator momentarily feels heavier upon moving upwards and feels lighter upon moving downwards, he noted that there was no intrinsic difference between *inertia* and

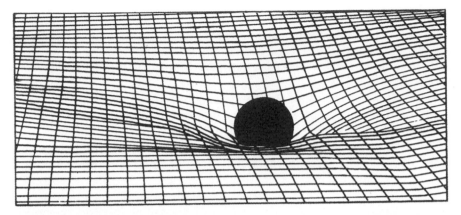

If space is represented by a rubber sheet inscribed with a grid, this depicts how it is distorted by the presence of a large mass.

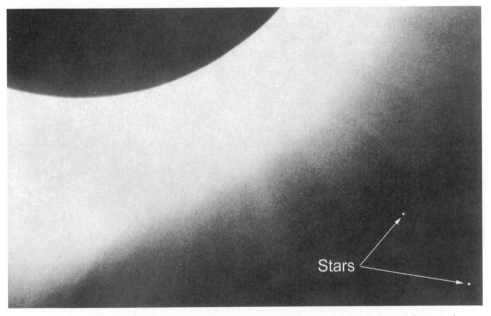

The prediction of general relativity that a ray of light would be deflected by passing close to a large mass was confirmed by measuring the positions of the stars in this photograph of the sky near the solar limb during the eclipse of 29 May 1919.

gravitation; that is, it was impossible to distinguish between a constant gravitational field and a uniformly accelerated frame of reference that was free of gravity. He formalised this as the Principle of Equivalence, and in 1911 decided to try to develop a full theory of gravitation. In 1915, by which time he had relocated to the Kaiser Wilhelm Physical Institute in Berlin, he had realised that gravitation was *a distortion of space* due to the presence of mass, and that the motion of masses was dictated by the manner in which space was curved. The traditional analogy involves a rubberised sheet: the sheet will be distorted when a heavy mass is placed upon it, and smaller masses that are placed near to it will tend to roll into the depression. In the abstract sense, the manner in which one mass 'attracts' another is indistinguishable from gravitation.

Einstein's General Theory of Relativity made specific predictions. In Newton's view, gravitation was a property of masses; however, Einstein had established that mass and energy were related, and his gravitation was a property of mass-energy. One prediction of the General Theory of Relativity was that a ray of light would be deflected when passing through space that was distorted. A.S. Eddington at the University of Cambridge was eager to make this test and, with the enthusiastic support of Frank Dyson, the Astronomer Royal, dispatched expeditions to Brazil and to the island of Principe off the west African coast to photograph the positions of stars close to the solar disk during the eclipse of 29 May 1919. The results confirmed the prediction.

The concept of curved space was not new. The term 'non-Euclidean' geometry was coined in Germany by J.K.F. Gauss, the greatest of the late eighteenth-century mathematicians. Although Gauss had 'dared' to question the teaching of the classical geometer Euclid, he kept his musings secret. The first comprehensive non-Euclidean geometry was developed independently by N.I. Lobachevski at the University of Kazan in Russia in 1829, and by the Hungarian engineer János Bolyai in 1832. In 1854, three years after receiving his doctorate at the University of Göttingen under the aged Gauss's direction, G.F.B. Riemann, generalised this for curvature in an arbitrary number of dimensions. Whereas Euclid had thought it axiomatic that *one and only one* straight line could be drawn through a pair of points, Riemann argued that, for a point *not* on a line, *no* line could be drawn parallel to that line; and any number of straight lines could be drawn though two points in Riemannian geometry. In a curved space of this type, the sum of the angles of a triangle always *exceeded* 180 degrees. Furthermore, there could be no such thing as a straight line of infinite length. The shortest distance between two points was the shorter of the two arcs of the 'great circle' that passed through them. In essence, while Euclidean geometry dealt with a plane, Riemannian geometry was that of a sphere. The British mathematician W.K. Clifford translated Riemann's work and in presenting it to the Cambridge Philosophical Society in 1870 suggested that the aether was a manifestation of the curvature of space. Karl Pearson exploited the fact that Riemann had generalised geometry to arbitrary dimensions, and in 1892 he proposed a hydrodynamical theory in which the aether entered 'our' space through 'sources' and departed through 'sinks' connected to another (fourth) dimension of space. When Einstein realised that one way of interpreting the Principle of Equivalence was to *ascribe* gravitation to the curvature of space, he was delighted to find that Riemann had already developed a suitable mathematical formalism.

In effect, Newton's gravity was a low-energy *approximation* to general relativity, because the differences became evident only for objects moving at high speeds or in deep 'gravity wells'. However, while Newton had had to discover the inverse square law empirically and insert it into his equation 'by hand', the relativistic formulation *required* it, which made it a more compelling theory. Also, interpreting gravitation as curved space banished Newton's concept of 'action at a distance'.

THE DISCOVERY OF THE ELECTRON

After a range of experiments by various people in Europe, in 1803 John Dalton in England argued that if there were different chemical 'elements', then each chemical 'compound' could be explained in terms of unions of small numbers of atoms (using Democritus's term). In his 1808 book *A New System of Chemical Philosophy,* Dalton chemically analysed all substances in terms of the 26 elements known at that time, and promoted his 'atomic theory' in which the atoms of the various elements – although distinct from each other – were still thought to be fundamental units of matter. W.H. Wollaston in England discovered in 1802 that several thin dark lines appeared in the solar spectrum if a narrow slit was placed in front of a prism, but he thought that they 'divided' the colours. In 1814, J. Fraunhofer in Munich resolved hundreds of lines utilising one of his own achromatic refractors, but their origin remained a mystery until 1859, at which time G.R. Kirchhoff and R.W. Bunsen of the University of Heidelberg noticed that the dark lines in the solar spectrum matched bright lines in the spectra of incandescent gases. J.J. Balmer, a Swiss schoolteacher with a passion for physics, set out to study glowing hydrogen because the arrangement of its spectral lines was simplest, and in 1885 he noted that the lines became more and more crowded with decreasing wavelength, with a spacing related to the square integers 1, 4, 9, 16 and so on. While this did not explain *why* the wavelengths were organised in this way, it gave an empirical means of calculating them, which was in itself a great advance. The existence of this 'Balmer series' suggested that there was a *structure* within the hydrogen atom – it was *not* an indivisible unit of matter.

Michael Faraday was frustrated in his efforts to determine if an electric discharge could propagate through vacuum because he could not make a satisfactory vacuum using the apparatus developed by Otto von Guericke. This limitation was eliminated in 1854 when Heinrich Geissler in Germany developed an efficient vacuum pump. In 1858 Julius Plücker applied an electric current to an evacuated glass tube and noted that a green glow, which ran between the cathode and the anode, formed inside the apparatus. Eugen Goldstein in Berlin suggested that this was a form of electromagnetic radiation emitted by the cathode, and in 1876 he introduced the term 'cathode ray'. William Crookes in England discovered in 1879 that such a ray could be deflected by a magnet, indicating that it was a stream of electrically charged corpuscles, and he speculated that the ray comprised electrified molecules, which he described as constituting a "fourth state of matter" (after solid, liquid and gas, and subsequently referred to as 'plasma').

In 1892 H.R. Hertz in Berlin tried to use an electric field to deflect the putative stream of negatively charged corpuscles, and upon seeing no deflection he concluded that the ray was electromagnetic. When his student, P.E.A Lenard, found two years

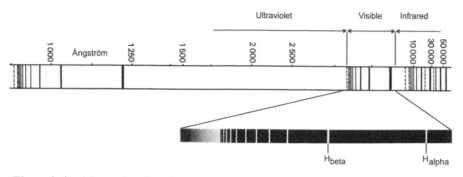

The emission/absorption lines in the spectrum of hydrogen extend to either side of the visual region.

later that a cathode ray could pass through a thin foil of metal, it was concluded that since atoms were believed to be spherical in shape and densely packed, with no room for matter to slip through, the ray must be some sort of 'flow' in the aether itself. However, J.B. Perrin in France established in 1895 that a ray could transfer a negative charge onto a metal block set in its path, and must therefore be a stream of charged corpuscles. The contradictory evidence was frustrating. In 1881 H.L.F. Helmholtz in Germany had suggested that "atoms of electricity", each of which had Faraday's unit charge, accompanied ions in an electrolysis solution. In 1891 G.J. Stoney in Ireland coined the term 'electron' for the entity carrying the unit of charge. In the early 1890s, H.A. Lorentz at Leiden University and Joseph Larmor in England both came to the conclusion that the electron was a manifestation of the aether, independent of matter. Stoney had suggested that electrons rotating in molecules might be responsible for their spectral lines. In 1895 Larmor suggested that every molecule incorporated a stable configuration of revolving electrons. Pieter Zeeman, one of Lorentz's research students, discovered in 1896 that a magnetic field 'split' spectral lines, and when he investigated this mathematically he found that the ratio of charge-to-mass was at least 1,000 times greater than that measured for ions by electrolysis, which implied that the mass of the electron was *tiny*. While ions can be either positively or negatively charged, a polarisation study revealed that all electrons were *negatively* charged. The reason for this was a mystery.

J.J. Thomson in England re-ran Hertz's experiment in 1897, applying a stronger electric field, and observed that the cathode ray *was* deflected. The degree to which a charged particle can be deflected by an electric field depends on its speed, its mass and the magnitude of its electric charge. The deflection by a magnetic field depends upon these same parameters but *differently*. Determining both deflections enables the charge-to-mass ratio of the corpuscles to be calculated. However, the first task was to measure their speed. Thomson installed into his evacuated tube a pair of plates athwart the beam in line with the magnets, and when the magnetic field deflected the beam in one direction he put an electric potential across the plates to deflect it back. For a specific magnetic field, the electric field that cancelled the

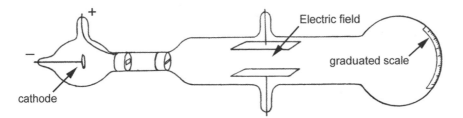

In investigating the nature of 'cathode rays' in 1897, J.J. Thomson placed plates for an electric field inside an evacuated tube in order to be able to control the deflection of the beam.

deflection enabled the speed of the particles to be calculated. At 200 kilometres per second the corpuscles were travelling at less than 0.1 per cent of the speed of light. A cathode ray was therefore not an electromagnetic wave. The only electrically charged entities known were ions (but, as yet, no-one was certain what they were). When Thomson calculated the charge-to-mass ratio, it confirmed Zeeman's inference that it was considerably greater than that of ions. To follow up, Thomson showed that this ratio was independent of the chemical composition of the cathode and the rarefied gas in the tube. Using Faraday's unit of electric charge, Thomson calculated that the mass of the corpuscles was minuscule – barely 1/1,000th the mass of the hydrogen atom, the lightest element. He realised that Lenard had seen a cathode ray pass through a thin film of metal because the tiny corpuscles could readily squeeze through the gaps between the atoms. Rather than pursue the line of thought that the corpuscles were a primitive manifestation of the aether, he saw them as evidence that (as Balmer had inferred) atoms were *not indivisible*. The cathode's intense electric field 'broke up' the atoms of the gas. The fact that the same negatively charged corpuscles were produced irrespective of the composition of the gas indicated that they formed *a universal subatomic constituent of matter*. When FitzGerald read Thomson's report he suggested that cathode rays were composed of 'free electrons', although Thomson had not used that term. As it had already been determined that an electric current was carried by negatively charged corpuscles, it was finally realised that cathode ray corpuscles were the *same thing* – the cathode ray was an electric current flowing through the tenuous gas in the evacuated tube.

When S.A. Arrhenius in Sweden submitted his doctorate in 1884 on the topic of ions, arguing that they were 'electrified atoms', his thesis referees – who believed that atoms were indivisible and therefore *could not* carry electric charges – awarded his degree the lowest possible grade. However, the discovery that the electron was a subatomic particle meant that an atom could acquire an overall charge if it could gain or lose an electron relative to its 'proper' complement. Arrhenius had the last laugh when his insight was rewarded in 1903 with the Nobel Prize.

SEEING THE LIGHT

When Max Planck entered the University of Munich in 1875 to study physics his professor warned him that the subject was essentially 'finished'. The renowned American experimental physicist A.A. Michelson ventured in 1894 that "it seems probable that most of the grand underlying principles have been firmly established", and in 1902 in his book *Light Waves and Their Uses* he wrote of a convergence of the physical sciences in which

> "the nature of atoms, and the forces called into play in their chemical union; the interactions between these atoms ... as manifested in the phenomena of light and electricity; the structures of the molecules and molecular systems of which the atoms are units; the explanation of cohesion, elasticity, and gravitation – all these will be marshalled into a single compact and consistent body of scientific knowledge."

Nevertheless, several puzzling observations had been made over the years. Kirchhoff realised in 1860 that incandescent gases emit at the *same* wavelengths as they absorb when cold. The colour of an object therefore depends on which wavelengths of sunlight it absorbs and which it reflects. Kirchhoff postulated that a body that absorbed at *all* wavelengths would appear perfectly *black*, and after being heated would radiate the energy it had absorbed by emitting it across the entire spectrum. While studying thermodynamics in 1879, Josef Stefan in Vienna noted a relationship between the fourth power of an object's temperature and the amount of energy it radiated. A 'black body' can therefore be defined as an object that (after being heated) emits all of the absorbed radiation in a spectrum that can be described *solely* by a *characteristic temperature*. Although most bodies reflect only a fraction of the incident radiation, physicists found a black body to be a useful mathematical concept. In particular, Wilhelm Wien, who studied under Helmholtz, conducted an experiment in 1893 in which he made an approximation to a black body in the shape of a cavity possessing a small hole in its wall. Any radiation entering the cavity through the hole should be absorbed and, in line with Kirchhoff's postulation, the radiation that subsequently emerged should have a black body spectrum. Upon measuring the spectrum of this radiation, Wien found firstly that the energy *peaked* at a certain wavelength, and that the wavelength of the emission peak was inversely related to temperature – that is, moderately hot bodies radiate mostly in the infrared, and when they are heated their peak of emission moves through the red end of the visible spectrum until they glow 'white hot'.

Wien soon devised a formula that related to wavelengths shortward of the peak, but failed beyond. In 1873 J.W. Strutt inherited the title of Lord Rayleigh, and a few years later succeeded Maxwell in the directorship of the Cavendish Laboratory in Cambridge. In 1898, working with recent graduate J.H. Jeans, he devised a formula from general principles of thermodynamics in which the amount of energy radiated by a black body varied inversely with the fourth power of the wavelength. Although this was a fair description of the spectrum at long wavelengths, the short-wavelength

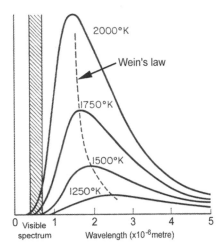

In 1893 Wilhelm Wien discovered that the emission from a 'black body' *peaked* at a certain wavelength that was inversely related to temperature.

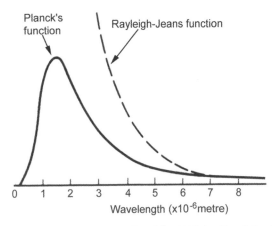

The energy spectrum of a 'black body' predicted in 1898 by Lord Rayleigh and J.H. Jeans failed at short wavelengths. When Max Planck introduced the idea of 'quanta' a few years later, he was able to account for the observed peak.

vibrations favoured by the model predicted that an *infinite* amount of energy would be radiated at exceedingly short wavelengths, which was clearly not the case.

In 1900 Planck set out to reconcile the conflicting formulas, one of which fitted shortward of the peak in the black body curve while the other fitted longward of it. In a remarkable insight, Planck realised that there was a flaw in the assumption that radiation was emitted *continuously*. It is absorbed or emitted in discrete units, which he called 'quanta'. In 'classical' physics, the energy of a wave is dependent upon its amplitude but *not* upon its wavelength, but Planck realised that it is independent of amplitude and dependent *only* upon wavelength. In fact, the energy of a quantum is inversely proportional to its wavelength. Specifically, $E = h\nu$. The factor h is known as Planck's constant. The frequency, ν, is inversely proportional to the wavelength. The long wavelengths tend to fit the Rayleigh–Jeans formula but the flood of short-wavelength vibrations that this predicted simply cannot arise. The spectrum reaches a peak and then tails off again and, as Wien established, the wavelength of this peak depends only on temperature.

This was a major advance, and it explained another puzzling observation that had defied explanation. The fact that light liberates electrons from the atoms in a metal plate was noted by Hertz in 1887. In 1902 Lenard observed a 'threshold' to this 'photoelectric effect'. At a wavelength longer than the threshold, no electrons were released irrespective of the intensity of the light. Electrons began to be emitted at the threshold but they were slow, and increasing the intensity simply increased their number, not their speed. At very short wavelengths the electrons emerged at much faster speeds. The mystery deepened when it was realised that the threshold was different for different materials. In 1905 Albert Einstein realised that Planck's quanta explained this threshold. In classical physics, the electrons in the metal atoms ought to be 'buffeted' by light waves in the manner of a tethered buoy in a storm, and whether or not they broke loose depended on the wave's amplitude, *not* its frequency. However, precisely the opposite was observed. Classical physics also predicted that *all* of the electrons near the surface of the metal would be excited, which was not the case. Einstein reasoned that only an electron that was 'hit' by a quantum of light would be affected, and it would be knocked out of the atom only if the energy of the quantum exceeded a specific threshold, and since Planck had shown that a quantum's energy was proportional to its frequency there ought to be a frequency threshold. *Nothing* should happen below this threshold. It was not that the electrons were excited to very nearly the point of escape, they were *indifferent* to such illumination. They could not escape by absorbing two lesser quanta. Above the threshold, the excess energy manifested itself as kinetic energy, so electrons liberated by very short-wavelength light emerged travelling more rapidly than those released by radiation just above the threshold. The intensity of the light was not proportional to the amplitude of the oscillations in the electromagnetic field, it was determined by the *number* of quanta. Increasing the intensity merely released *more* electrons. This precisely explained Lenard's observations. It was this explanation of a phenomenon that had hitherto appeared to defy the laws of physics that convinced sceptics that although light was an electromagnetic wave, in certain circumstances it behaved as if it was corpuscular. This was significant as it eliminated the *requirement* for the aether: while it was difficult to imagine how a wave could travel though a vacuum, a particle would have no such difficulty. This drove the final 'nail in the coffin' of the concept of the aether.

3

The structure of the atom

MYSTERIOUS RAYS

The discovery of the electron as a universal subatomic particle undermined the ancient idea of atoms as the fundamental units of matter. The first clues regarding the structure of the atom came from mysterious 'rays'.

While H. R. Hertz and J. J. Thomson were independently pondering the nature of cathode rays, in 1895 W.K. Röntgen at the University of Würzburg investigated their effects on fluorescent substances. A screen of fluorescent material just beyond the end of the cathode ray tube glowed when the tube was active. For better observation he blacked out his laboratory and noticed that *another* screen, not on the axis of the tube, also fluoresced. In fact, so did one in the next room! Evidently, the tube was emitting a ray in addition to the familiar (but still mysterious) cathode ray which had sufficient penetrating power to pass through the wall. Since it defied explanation by established principles, he named this the 'X-ray'. An intense debate followed concerning whether they were waves or corpuscles. Part of the difficulty in determining their nature was that different experiments gave support to different interpretations. Röntgen noted that X-rays displayed some, but not all, of the characteristics of light and speculated that the rays might be longitudinal (as opposed to transverse) vibrations in the aether. However, in 1905 C.G. Barkla at the University of Liverpool inferred from polarisation observations that they were electromagnetic waves. By 1910 it was generally presumed that X-rays were very short-wavelength waves and, if this was true, they ought to be susceptible to diffraction. The difficulty in testing this hypothesis was the impracticability of making a sufficiently fine diffraction grating. M.T.F. Laue at the University of Munich suggested that the spacing of the layers of some crystals should be appropriate, and this was verified in 1912 by his colleague Walter Friedrich and his student Paul Knipping. The following year W.H. Bragg at the University of Leeds, and his son W.L. Bragg who had recently graduated from Cambridge, used diffraction to determine that the wavelengths of X-rays extended shortward of ultraviolet to about 10^{-11} metre. As energy is inversely proportional to wavelength, X-rays are extremely energetic, and hence penetrating. The fact that X-rays emerged from atoms implied that atomic structure involved tremendous energies.

On 25 February 1896, barely two months after Röntgen reported his discovery of X-rays, A.H. Becquerel – a physicist at the Museum of Natural History in Paris who, with his father A.E. Becquerel, had made a study of light-induced fluorescence – determined to find whether a fluorescent salt of uranium emitted X-rays. He sealed a photographic plate in light-proof paper, placed a metal cross on it, applied a second layer of wrapping and sprinkled the salt over it. When exposed to sunlight, the salt absorbed light and re-emitted the energy at its characteristic fluorescent wavelength. Although the fluorescence could not penetrate the paper shield, the plate was fogged with an impression of the cross, so Becquerel presumed that the salt had emitted X-rays. He made a second package to repeat the experiment, but the weather was poor and he put the package in a drawer to await a sunny day. A week later he developed the plate to verify the integrity of the light-proofing and found it to have been fogged even though the salt had *not* been able to fluoresce. He concluded that the emissions must not have been X-rays, but some new kind of ray. Further tests established that non-fluorescing uranium salts emitted similarly, as indeed did metallic uranium. How a material could emit even when *not* excited by a cathode ray or by sunlight was an even deeper mystery than how X-rays originated. Becquerel's 'uranium ray' did not initially attract as much interest as X-rays because it was much less intense. However, Marie and Pierre Curie, a husband and wife team working in Paris, found in 1898 that thorium emitted such rays strongly, and later that year Mme Curie found radium, which was a million times more potent than uranium. Since this was a general phenomenon, she introduced the term 'radioactivity'. A 'false start' in explaining this behaviour was the presumption that *all* matter was radioactive to some degree, but as instruments were improved this was shown to be false. This realisation in turn prompted the question of why *only some* materials behaved in this manner. In 1896, Gustave Le Bon, a psychologist by profession and physicist by inclination who believed matter to be a structure of the aether, had opined that it was the spontaneous decay of matter back into aether. In 1898 Ernest Rutherford, a young New Zealander, was appointed professor of physics at McGill University in Montreal, Canada. In 1899 he discovered that there were *several* such rays. One type that was easily obstructed, he named 'alpha'; another that was able to penetrate significantly, he named 'beta'. In 1900 Becquerel found that beta rays were readily deflected by a magnetic field, and were therefore negatively charged particles. In fact, they were identical to cathode rays, and hence were high-speed electrons. Something (evidently not the photoelectric effect) was ejecting electrons from atoms. When P.U. Villard in France identified even more penetrating 'gamma rays' in 1900, Rutherford argued that they were electromagnetic waves, and this was proved in 1912 when their wavelengths were shown to be considerably shorter than X-rays.

When, in 1902, Rutherford and Frederick Soddy discovered that thorium gave off an inert gas, they realised that this was not the result of a *chemical* reaction, but was a by-product of elemental *transmutation*. In 1903 they identified the gas given off by uranium as helium, which was another inert gas. Although it was obvious that

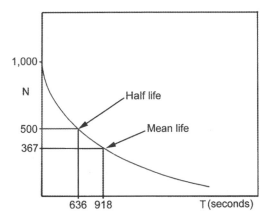

The radioactive half life is the time required for half of the atoms in any given sample to decay. The mean life, or lifetime, is the time for the number to be reduced by two-thirds (specifically, by a factor that is the reciprocal of 2.72, the base of the natural logarithmic scale).

radioactivity was due to a sudden alteration of the internal configuration of an atom, virtually nothing was known concerning the structure of the atom. In 1903 Pierre Curie found that a single gram of radium could heat its own weight of water from 0 °C to 100 °C in only an hour, confirming the suspicion that atoms contained tremendous energies. In 1904 Rutherford realised that the intensity of radioactivity decreased over time for a specific sample of material. As this 'decay constant' was specific to each isotope, it was possible to measure its 'half life' and its 'mean life' or 'lifetime'. As the process occurred spontaneously, the probability of any given atom decaying was *independent of its age*. While it was not immediately acknowledged, the statistical character of the decay process meant that radioactivity was causally unpredictable, even in principle.

THE NUCLEUS

If there were negatively charged electrons inside atoms, the fact that atoms were electrically neutral implied that there must be a matching positive charge, so in 1898 J.J. Thomson proposed that the bulk of an atom's spherical volume was a positively charged mass in which the electrons were embedded. As an analogy, he suggested the plums in a pudding, thereby ensuring that it became known as the 'plumb pudding' model. In 1903 P.E.A. Lenard concluded that the traditional view of atoms as bulky spheres was untenable, and suggested instead that an atom comprised a *cloud* of subatomic particles, some negatively charged and others positively charged, with their numbers in equal proportion. With the cloud of particles swirling around a common locus, the atom would effectively occupy a spherical volume but be mostly empty space. In retrospect, therefore, it was hardly surprising that a thin film of metal was permeable to cathode rays. Although Lenard had no idea what the positively charged particles might be, they were clearly not a positively charged form of the electron because the mass of the electron was tiny compared to even the lightest of atoms. However, in 1899 C.T.R. Wilson and John Townsend at the Cavendish Laboratory had shown that the hydrogen ion had precisely the opposite electrical charge to the electron and so it began to look as if the positively charged particles within atoms were hydrogen ions. Hantaro Nagaoka in Japan suggested in 1904 that the positively charged part of an atom was concentrated into a *nucleus* (meaning 'kernel' in Latin) which, although much smaller than the atom's volume, nevertheless contained almost all of its mass. In Nagaoka's 'nuclear' model of the atom, the cloud of electrons revolve around the nucleus in much the same way as the planets orbit the Sun, but with electromagnetism instead of gravitation providing the attractive force. When early tests with a magnetic field failed to deflect the alpha ray of radioactivity, it was concluded that it was electrically neutral, but in 1905 Ernest Rutherford discovered that its corpuscles carried an electric charge of $+2$ and were so massive that they would not have been significantly influenced by the weak field that readily deflected an electron. In 1908 Rutherford accepted a post at Manchester University in England. Soon thereafter he found that the alpha particles were similar in mass to the helium atom, and the following year a spectroscopic analysis showed that they were *identical* to helium ions. Since they emerged from radioactive atoms at such high speed, he decided to use them as 'projectiles' to investigate the structure of the atom. He placed a source of alpha particles inside a box with a small hole in one wall and measured how the narrow beam of alpha particles leaking out of the hole was scattered upon striking a thin film of metal placed just beyond: most of the alpha particles passed straight through, many were deflected (in some cases by substantial angles) and one in a thousand came right back.

It is not clear whether Rutherford had heard of Nagaoka's idea of a nucleus, but by 1911 he had established from the manner in which the positively charged nuclei deflected the trajectories of the similarly charged alpha particles that 99.9 per cent (even more in the case of the heavier elements) of an atom's mass was concentrated in a tiny nucleus that represented just 10^{-5} of the radius of the atom, showing that atoms really were mostly empty space. Although Nagaoka was the first to have suggested the idea of the nucleus, he had only been speculating and the idea had

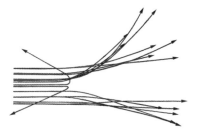

In 1908 Ernest Rutherford fired alpha particles at a thin film of metal and observed that many were deflected (in some cases by substantial angles) and one in a thousand came back.

languished. It was Rutherford's ingenious experiment that provided the evidence and the physical measurements. In 1914 Rutherford suggested that because the simplest nucleus, the hydrogen ion, was a subatomic particle, it should be named the 'proton' (meaning 'first' in Greek, and reflecting the fact that it forms the nucleus of the first element in the Periodic Table). It remained a mystery, however, why particles of equal and opposite electrical charges should have such different masses. As the electron appeared to be a 'secondary' component of an atom in the sense that it was so lightweight and peripheral to the nucleus, it was speculated that the nucleus was the 'real' spherical solid ultimate unit of matter – a sort of 'atom within an atom'.

BOHR'S ATOM

In 1900 J.R. Rydberg in Sweden empirically derived the relationship between the wavelengths of the emission lines in the spectrum of hydrogen that J.J. Balmer had noted in 1885, and by providing a scale in the form of a constant of proportionality (dubbed the 'Rydberg constant') he gave physical substance to the otherwise *ad hoc* Balmer series. In terms of Thomson's 'plum pudding' model, it was expected that the electrons embedded in the voluminous positively charged 'pudding' would be oscillating in a manner imposed by the distribution of positive charge, but the details were obscure. Rutherford's demonstration that there was no 'pudding' rendered this theorising obsolete and provided a 'clean sheet' on which to develop a totally new approach for the production of spectral lines. As hydrogen comprised only a single proton and a single electron, it served as a benchmark for testing theories of atomic structure. The analogy of electrons orbiting like planets was appealing, but as a planet pursues its orbit its velocity vector sweeps out 360 degrees and, as a result of this continuous acceleration, it radiates gravitational energy and slowly spirals in. The force binding an electron to its nucleus is electrostatic rather than gravitational, but the laws of classical physics would oblige an electron to radiate at a wavelength corresponding to the rate at which it was shedding energy. This was worrying for several reasons. First, it would result in a migration of the emission towards ever-shorter wavelengths. The electrons in recently created atoms ought to be far from the nucleus and lose energy slowly, whereas those in older atoms would be close in and lose energy rapidly. In fact, physicists had been wondering why *all* hydrogen atoms radiate using the *same* series of wavelengths. Even more alarming was that ultimately *all* electrons would spiral in and coalesce with their nuclei. Clearly something was amiss. After a brief visit to Cambridge, the young Danish physicist Neils Bohr commenced his apprenticeship in Rutherford's laboratory in Manchester. In addressing this mystery head-on, Bohr decided that if electrons ought to spiral in but did not do so, then the task was to discover *why* they did not do so. What, he wondered, could prevent electrons from continuously emitting electromagnetic radiation. Given the Balmer series, he postulated that electrons were restricted to 'stable' orbits and could absorb or emit at only certain wavelengths whose energies corresponded to the *differences* between pairs of stable orbits. As Planck's constant is actually a measure of angular momentum, Bohr suggested that the angular momentum of an electron within an atom was also quantised, which would limit the number of circular orbits available to those with an angular momentum that was a whole multiple of Planck's constant. He also reasoned that because an electron could not spiral from one orbit to another, it must 'jump'. Although such an essentially instantaneous transition was impossible by classical physics, it was plausible if the energy difference between the orbits was absorbed or emitted as a single quantum. Upon calculating that the wavelengths of the 'transitions' for hydrogen closely matched the Balmer series, Bohr labelled a series of integers in his formulation the 'quantum number' of an electron's state – in effect these numbers (which were whole number units of angular momentum) set the radii of the stable orbits, and the transitions became known as 'quantum jumps'. Bohr's insight explained (a) why electrons did not continuously radiate, (b) why *all* the atoms of a given element display the *same* set of characteristic spectral lines and (c) by establishing a 'ground state', he explained why electrons did not coalesce with the nucleus. However, it was not evident why the electrons in any given atom did not all drop down into this ground state.

In 1916 A.J.W. Sommerfeld in Germany noted that Bohr had presumed *circular* orbits, and he investigated what would happen if the electrons traced out *elliptical* orbits (as do the planets). By introducing a second integer series, which he called the 'orbital quantum number', he found that each spectral line in Bohr's model became a closely spaced set of lines. Bohr and Sommerfeld had both assumed that the orbits would all lie in the same plane (as do the planets, more or less). However, an electron will generate a magnetic field as it travels its orbit, and if its atom sat in an ambient magnetic field some of the stable orbits would align themselves at certain angles, so a 'magnetic quantum number' was added. In the absence of an ambient field, electrons differing only in this quantum number would have the same energy. In 1924, the Austro-Swiss physicist Wolfgang Pauli, once a student of Sommerfeld and Bohr, said that the fine line-splitting detected by Pieter Zeeman in 1896 required a fourth quantum number, and G.E. Uhlenbeck and S.A. Goudsmit in Holland suggested that if electrons rotated axially (as do planets) they would generate a small magnetic field that could align itself with or against a prevailing magnetic field. In order to define these states as being one unit apart, the 'spin quantum number' was restricted to the values $\pm 1/2$. In the presence of an ambient field, each spectral line predicted by the state of the first three quantum numbers would be split in two. Pauli also discovered that no two electrons in any given atom could have the same 'quantum state', this being defined by the entire set of quantum numbers. The rule, which he described as the *Exclusion Principle*, explained why the electrons in an atom do not all drop down into the ground state.

Quantum numbers were empirical, but the fact that the formalism utilising them described the hydrogen spectrum in exquisite detail hinted at a fundamental structure within the atom. Nevertheless, Bohr's model envisaged the electron as a *particle* that pursued its orbit around the nucleus as a planet orbits the Sun, and this was merely an analogy. What was the electron, really? In 1925 W.K. Heisenberg, recuperating on the island of Heliogoland from an attack of fever, rejected Bohr's model, arguing that there was no evidence that the parameters used had any physical

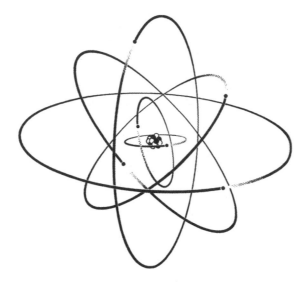

In the Rutherford–Bohr atom, the electrons travel in orbits at various inclinations around a dense nucleus. This schematic representation was adopted as its logo by the US Atomic Energy Commission.

significance. Such a mechanical model, he said, was an over-interpretation of a few basic observations. It was possible to see light of different wavelengths shining within an atom, but it was not possible to see electrons in orbits. In an effort "to destroy without trace the idea of orbits", Heisenberg devised a theory using *only* parameters that were observable and whose values could be derived by experiment. As he represented the parameters of his mathematical formalism employing matrices, the theory for describing spectra independently of an interpretive model was called 'matrix mechanics'. It worked, but was it any more valid than Bohr's approach? In fact, there was already evidence that undermined the idea of orbits. A.H. Compton in America had realised in 1923 that a quantum of light could undergo an 'elastic' collision (that is, in the manner of a ball on a billiards' table) in which it transferred *some of* its energy to the electron and left with the residual energy in the form of a quantum of a longer wavelength. After Planck had revealed quanta to be absorbed or emitted whole, this particle-like scattering (soon dubbed the 'Compton effect') came as a considerable surprise. In 1926 G.N. Lewis proposed, and Compton agreed, that the electromagnetic quantum be referred to as a 'photon' (meaning 'light' in Greek).

The debate, since Newton and Huygens, on the nature of light was belatedly realised to have been futile, because light displays *both* wave-like and particle-like behaviours. Intriguingly, its behaviour is dependent upon the *manner* in which it is observed. In general, if an experiment is arranged to detect characteristics associated with waves, wave-like behaviour will be seen; and if it is set up to detect particle-like behaviour, then particle-like behaviour will be seen. Furthermore, behaviour in the manner of a particle is more readily evident towards shorter wavelengths, and is

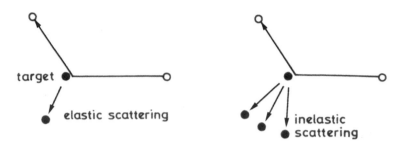

In an elastic collision, the impactor's kinetic energy is divided with the target, but in an inelastic collision the target is disrupted and the energy is shared out. Whereas in an elastic collision the sum of the kinetic energy is the same after as before, it is not so in an inelastic collision because energy is consumed in breaking up the target.

particularly noticeable in the case of X-rays and gamma rays. In fact, this *duality* indicates that electromagnetic radiation is *neither* a wave nor a particle, but is something more fundamental that can display both types of behaviour depending on the circumstances. In his doctoral thesis in Paris in 1924, L.V. de Broglie suggested that this duality was *universal*. In particular, he realised that combining Einstein's $E = mc^2$ and Planck's $E = hv$ implied that the electron *should* display wave-like behaviour. In 1925 G.P. Thomson (son of J.J. Thomson) passed a beam of electrons through a gold foil, and confirmed the predicted diffraction effect. At the same time, C.J. Davisson and Lester Germer of the Bell Telephone Laboratories in America found that electrons were diffracted by a crystal and produced the interference pattern on a phosphorescent screen which was identical to that expected of a wave. In effect, this reproduced Young's 'double slit' experiment, in this case using electrons instead of light.

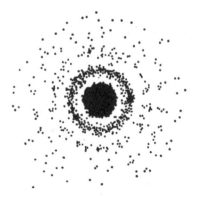

In 1925 G.P. Thomson passed a beam of electrons through a gold foil and inferred from the resulting diffraction pattern that they exhibited wave-like properties.

QUANTUM MECHANICS

The fact that electrons displayed *wave-like* behaviour caused Austrian theoretical physicist Erwin Schrödinger to realise in 1925 that that was *why* only certain orbits in an atom were stable – the electron's waveform would extend around the path but only certain radii would create 'standing waves', the waveform would be destroyed by self-interference at all other radii. This also explained why the electron *could not* coalesce with its nucleus. Its ground state was the closest that it could approach the nucleus because there was no way for it to lose further energy. This explained why atoms were stable. Schrödinger formulated a 'wave function' to describe the electron. Upon applying the resulting 'wave equation' to the hydrogen atom, it produced the Balmer series. The series of integers that Bohr had introduced to specify the number of units of angular momentum for the orbit was renamed the 'principal quantum number'.

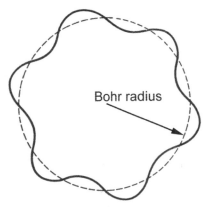

When the de Broglie wave for the electron in a hydrogen atom was calculated, its mean radius was found to correspond to that of the orbit in Bohr's model.

Schrödinger did not know what his oscillating wave function signified in terms of physical properties; he only knew that for an electron within an atom it had to be a stable configuration. At first, the wave function was thought to imply that the electron was physically smeared out, but in 1926 Max Born in Berlin realised that the square of the amplitude of the wave function was really the *probability* of the corresponding particle being at that particular point. In a sense, it did not matter where the electron was at any particular time, what mattered was the *distribution of its electric charge*, which was proportional to the probability of the electron being present at each point in the vicinity of the atom. Bohr's concept of electrons orbiting the nucleus was indeed an over-interpretation. The wave function of the electron defined zones of electric charge within the atom.

The mystery of the wave–particle duality was lifted by thinking of the electron as a *wave packet* in which the amplitudes of a succession of waves are greatest in the middle of the 'packet'. Particle-like behaviour arises because the amplitude at a point

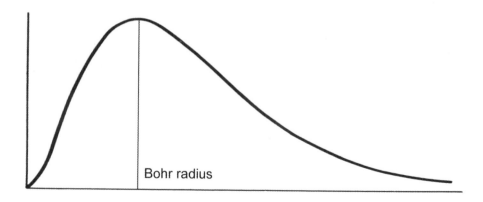

Bohr radius

When the square of the wave function was interpreted in terms of probability, the peak indicated that the most likely place for the electron was in the orbit in Bohr's model. In fact, because we cannot know where the electron is, the function profiles the distribution of its electric charge in the space surrounding the nucleus.

in space is initially zero, indicating that the particle is not present; as the packet sweeps by, the amplitude fluctuates, indicating that the particle is nearby, after which the amplitude returns to zero as the particle departs.

The concept of the wave packet provided insight into how a particle could exhibit both wave-like and particle-like behaviour.

Schrödinger's version of the wave equation dealt with energy and momentum in a classical manner, employing a first-order time-related derivative and a second-order space-related derivative. This was a problem, because subatomic particles travel at a significant fraction of the speed of light. Schrödinger recognised this, but was unable to incorporate relativity. There were two methods of eliminating this asymmetry to make a relativistic solution: (1) to modify the time-related term to match the space-related term, or (2) to modify the space-related term to match the time-related term. In 1926 Oskar Klein and Walter Gordon independently pursued the first option, and derived a solution resembling an electromagnetic wave with an additional term to accommodate the rest mass of the electron. However, because it did not incorporate the concept of spin (or rather, it described 'spinless' particles) it could not recreate the fine detail in the hydrogen spectrum. In 1927 P.A.M. Dirac in Cambridge set out to eliminate the asymmetry in the equation by the second, more difficult option.

All particles can be described as oscillations in associated fields. The concept of a field was developed by J.C. Maxwell in unifying electricity and magnetism. There are several different kinds of field. A *scalar* field – such as temperature – is described by a value representing the 'potential' of the field at each point in space. A *vector* field is described by a potential with a vector because the field incorporates a direction. The photon is an oscillation in Maxwell's electromagnetic field. When Einstein devised general relativity, he employed *tensor* fields to represent geometry. All these types of field are symmetric under rotation such that they yield the same value after a 360-degree cycle. Since turning everyday objects through a full rotation made no change, it was presumed that this must *always* be so, but Dirac realised that the electron required a field that did *not* obey this symmetry. Whereas time has only one dimension and space has three, he had to invent (what he called) a *spinor* field in which a 360-degree rotation just inverts the sign of the potential, and a 720-degree rotation is required to reinstate the initial value. He discovered that a particle represented by a spinor field possesses angular momentum even when it is at rest in space, and named this '*intrinsic spin*' (usually abbreviated to 'spin'). The 'up' and 'down' orientations of spin are defined relative to the observer's arbitrary frame of reference.

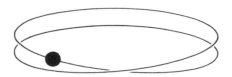

When a particle represented by a spinor field undergoes a 360-degree rotation, this merely inverts the sign of its potential; a 720-degree rotation is required to reinstate the initial value.

In Dirac's formulation, electrons differ from each other only in terms of *energy*, *momentum* and *spin* – all of which are properties characterising the way in which the *wave function* responds to *symmetry transformations*. The variation of the *amplitude* of the wave function in time is defined as its *phase*. Its energy is defined as the change in phase (in cycles, or parts of cycles) over an interval of 1 second, and expressed in units of Planck's constant. In much the same way, the component of its momentum in any particular direction is defined as the change in phase when the point from which positions are measured is adjusted by 1 metre in that direction, again in units of Planck's constant. The amount of spin around any axis is defined as the change in phase when the frame of reference for measuring directions around that axis is adjusted by 360 degrees, times Planck's constant – which is why the spin quantum number is given in 1/2-integer steps. In effect, therefore, Planck's constant converts from empirical macroscopic units of measure to the unit appropriate to the quantum realm. Energy is an intrinsic property of fields, but momentum and spin are extrinsic because of the *invariance* of the laws of nature to changes in the frame of reference utilised to measure positions and directions in space. It was this symmetry that made the wave equation consistent with relativity. By modifying the space-related term of the wave equation to match the time-related term, Dirac discovered that the quantum concept of spin is *not* a dynamical property, it is a manifestation of the geometry of space. Significantly, if the property of spin had not already been inferred empirically, Dirac would have been able to predict it.

The spinor explained a puzzling observation. If a particle had an electric charge, its intrinsic spin would generate a magnetic field. The magnetic field of the electron had been measured with precisely twice the strength predicted by the notion of a particle as a spinning ball. This was explained by the fact that the spinor must rotate twice to complete a phase cycle. In fact, this observation was considered to be proof that Dirac's equation was valid. In directly encompassing the wave–particle duality, this equation produced a wave-like behaviour when 'probed' in a wave-like manner and a particle-like behaviour when probed in a particle-like manner *without introducing a paradox*. The properties that could be observed in each situation initially seemed to be different, but there were correspondences. In particular, the intrinsic spin of the electron turned out to correspond to the *polarisation* of its wave form. In fact, the familiar phenomenon of polarised light is a manifestation of the photon's spin. If the spins of a large number of photons are randomly aligned, the light is unpolarised, but if the spins tend to be in one particular direction, we say that the light is polarised.

Dirac's formalism, which was published in January 1928 in a paper entitled 'The Quantum Theory of the Electron' addressed only the electron and electromagnetic force, so in 1929 Heisenberg and Pauli independently generalised it as *quantum field theory* with a specific field for each and every type of particle. In fact, Schrödinger's 'wave mechanics' and Heisenberg's matrix mechanics were revealed to be equivalent, and both were candidates for describing what was beginning to become known as 'quantum mechanics'. In terms of quantum fields, particles are concentrations of energy arising from field *oscillations* – for example, a light wave is composed of oscillating electric and magnetic fields, and the passage of the disturbance in the field manifests itself as the wave packet that we interpret as the particle called the photon.

UNCERTAINTY

In 1927, while studying under Bohr in Copenhagen – and after having used matrices to represent the parameters in his wave mechanics and observed that they violated the law of commutative multiplication – Heisenberg realised that this rendered it impossible, even in principle, to simultaneously determine with *exact* precision an electron's position and momentum. In fact, this can be expressed by the relationship $\Delta p \Delta x > h$, where h is Planck's constant. The time interval of an observation must be constrained to determine the position (x) of a particle, but must be extended long enough to measure its speed in order to calculate its momentum (p). Although these parameters *can* be measured to arbitrary precision, the uncertainty in the position (Δx) decreases as the uncertainty in the momentum (Δp) increases and it is not feasible to know *both* with fine accuracy. This limitation is not an issue of better instruments, it is fundamental to the quantum realm. When Heisenberg showed this analysis to Bohr, he realised that the conservation of mass-energy *required* it to be so.

By denying the possibility of determining both the position and momentum of a particle, the Uncertainty Principle (as it was called) rendered pointless the efforts to describe an atom in terms of electrons in orbits. It did not affect Schrödinger's wave function, however, because this described the electron in terms of the distribution of electric charge in the space around the nucleus without stating *where* the equivalent particle was.

FERMIONS AND BOSONS

In the quantum realm, particles cannot spin at arbitrary rates. The 'basic rotation rate' is derived by setting the angular momentum equal to Planck's constant, making allowance for the fact that heavier particles, being physically larger, rotate more slowly than smaller ones due to the difference in their effective radius. This effect can be readily observed when an ice skater extends or draws in his arms. The proton's basic rotational rate is 10^{22} times per second. Multiplying the basic rate by one of a set of spin factors 0, 1/2, 1, 3/2, 2, etc, allows for the rotations of particles of different masses. The angular momentum of a particle is therefore essentially the product of the spin factor and Planck's constant. Although protons and electrons both have the same spin factor of 1/2, a proton is more massive and will rotate more slowly.

In light of Dirac's discovery that a photon can be described by a vector field but an electron requires a spinor field, it became evident that the whole-integer and half-integer spin factors reflected a fundamental division into two classes of particle. The difference in the fields means that when two half-integer spin particles of the same type (for example, two electrons) are in such close proximity that their wave functions overlap, the probability of their occupying the same point is zero – this is the essence of Pauli's Exclusion Principle. However, if two whole-integer spin particles (for example, two photons) overlap, they do *not* cancel out and the probability of their being in the same energy state is non-zero.

The properties of particles with *integer spin* (0, 1, 2, etc) had been studied by S.N. Bose in India in 1924 and by Albert Einstein in 1925. Since at that time no 'spinless' (spin = 0) particles were known, this type consisted of the electromagnetic photon (spin = 1) and the hypothetical graviton (spin = 2). As Bose–Einstein statistics applies to systems of particles in which the individuals cannot be distinguished from each other, an unlimited number of photons can co-exist in the *same* energy state. Independent theoretical studies of electrons by Enrico Fermi in 1926 and Dirac the following year produced Fermi–Dirac statistics for a system of particles in which the individuals *can* be distinguished from each other. In fact, this applies to all particles having *half-integer spin*. At that time, electrons and protons were the only examples, but in 1945, by which time more were known, Dirac named the former group 'bosons' and the latter 'fermions'. Fermions obey the Exclusion Principle, but bosons do not. As the known fermions tended to possess mass and served as subatomic particles, while bosons such as the photon and graviton were massless and were the 'agents' of the electromagnetic and gravitational forces, it was realised that the difference in the spin properties of the vector boson fields and the spinor fermion

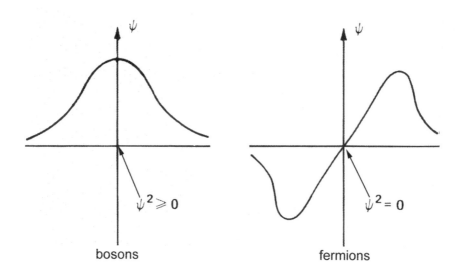

bosons   fermions

When two fermions of the same type are in such close proximity that their wave functions overlap, the probability of their occupying the same point is zero, but if two bosons overlap they do *not* cancel out and the probability of their being in the same state is non-zero.

fields was fundamental. If fermions in the same state were able to coexist then atomic matter would not exist because unlimited numbers of electrons would accumulate in the ground state of an atom. The essential character of our Universe therefore derives from this distinction.

INSIDE THE NUCLEUS

It was evident that the electrons liberated by the photoelectric effect came from the orbiting cloud and those ejected by radioactive decay originated from the nucleus. The alpha particles, which were indistinguishable from helium nuclei, could only have originated from the nucleus. The nucleus therefore had a structure of its own, and was not the suspected 'atom within an atom'. The gamma-ray radiation was too energetic to have been produced by electrons jumping between quantum states. If this radiation represented nuclear transitions, and as such was only the *difference* between nuclear states, then that tiny structure must embody a tremendous concentration of energy.

In the 1920s, it was presumed that the helium nucleus, weighing 4 atomic mass units and having an electric charge of $+2$, comprised four protons and two electrons, and it was presumed that this scheme scaled up to account for each of the 100 or so elements. Matter, it appeared, was composed entirely of two subatomic particles – the electron and the proton. This was a remarkable simplification over the previously daunting variety of Dmitri Mendeleev's Periodic Table of Elements. It also appeared that there were only two fundamental forces – electromagnetism and gravitation. While these seemed to be so different on a macroscopic scale, with electromagnetism being described in terms of quantum theory and gravitation in terms of the curvature of space, Albert Einstein was already trying to formulate a 'unified field theory' that would unite them in much the same way that J.C. Maxwell had unified the apparently disparate electric and magnetic forces.

In 1920 Ernest Rutherford pondered why (with the exception of hydrogen) the atomic mass of a nucleus was twice that of its electric charge, and wondered whether this might indicate that the electrons and protons in the nucleus *fused*, perhaps with an electron bound in a tight orbit around a partner proton, with each pair behaving as if it were a single electrically neutral particle with spin values –1, 0 and $+1$ depending on whether the $\pm 1/2$ spins of its constituents added or cancelled. Although the terms had not yet been invented, this would mean that when the fermions fused they created a boson. However, studies of the angular momentum of nuclei indicated that if such a neutral particle was present, it *must* be a fermion. This suggestion therefore seemed to violate the conservation of angular momentum. Rather than yield this law, physicists continued with the idea of a nucleus composed of protons and electrons. The problem was that the model was valid for even atomic numbers, but not for odd ones. Spectroscopic evidence indicated that the overall spin on a nucleus was ± 1. In the case of helium (with atomic number 2) two of the positive charges on the four protons were cancelled by electrons, the number of particles (six) was even, and the overall spin was whole-integer, but for lithium (with atomic number 3) three of the six protons were cancelled by electrons, the number of particles (nine) was odd, and the overall spin was *half-integer*. Resolving the dilemma meant either rejecting a conservation law or reliable spectroscopic evidence.

In 1930 W.W.G.F. Bothe in Germany discovered that bombarding beryllium foil with alpha particles produced an electrically neutral penetrating radiation which he presumed to be gamma rays. However, in 1932 in France, Frédéric Joliot-Curie and his wife Irène (daughter of Pierre and Marie Curie) found that this was *not* a gamma ray. In 1932 James Chadwick in Cambridge, a former student of Rutherford, studied this mysterious ray in a cloud chamber. Since the apparatus revealed the trails only of electrically charged particles the ray could not be seen entering the chamber, but if it passed close to an atom of the gas this was ionised and the track of the ion's recoil enabled the properties of the intruder to be determined. It was slightly heavier than the proton, and a fermion. This could only be the neutral particle that Rutherford had postulated, and in 1921 W.D. Harkins in America had suggested be named the 'neutron'. Heisenberg immediately pointed out that nuclei composed of protons and neutrons were *consistent* with the observed nuclear spin. The neutron therefore joined the proton and electron in the family of subatomic particles. It had taken so long to reveal itself because the easiest property to observe in a particle was its electric charge, and the presence of a neutral particle could only be inferred indirectly by its effect on charged particles.

The discovery of the neutron explained how the nuclei of the isotopes of the light elements hydrogen and helium were constructed.

ALPHA DECAY EXPLAINED

At the University of Göttingen in 1929, a recently graduated PhD, George Gamow, made the first successful application of the quantum theory to the mysteries of the atomic nucleus by working out how an alpha particle escapes from a nucleus during radioactive decay. Gamow realised that Heisenberg's Uncertainty Principle enabled a particle to 'tunnel' its way out. If the positively charged alpha particle was in a low-energy state within a large nucleus, it would be confined by the force that held the nucleus intact; but if it was excited, the momentary uncertainty in its energy meant that it stood a chance of straying beyond the nucleus, where it would be driven off by the repulsive electrostatic force.

NEUTRON INSTABILITY

If the nucleus was composed of protons and neutrons, where did the electrons of beta decay come from? In 1934 refined measurements established that, contrary to an early estimate, the neutron's mass was slightly *greater* than the sum of the masses of a proton and an electron, and in 1935 Chadwick and Maurice Goldhaber deduced that a *free* neutron would spontaneously decay into a proton and an electron, their electrical charges would cancel out, and the excess energy from the mass conversion would manifest itself as the kinetic energy that sent the particles racing away from one another. This process was not actually observed until 1948 by A.H. Snell at the Oak Ridge nuclear facility in Tennessee. Once statistics were available, J.M. Robson in Canada calculated in 1950 that the half life of a free neutron was 13 minutes. However, since this was a statistical measure, it was impossible to predict when any particular neutron would decay.

At first, considering the role that the neutron plays in the nucleus, the instability of the free neutron posed a quandry. In fact, neutrons in the nucleus are not free. In a stable nucleus, the neutrons are *inhibited* from decaying by the fact that the protons occupy a set of energy states that are subject to Pauli's Exclusion Principle and are so tightly packed that, quite simply, the nucleus cannot accommodate another proton. In an unstable nucleus, where there *is* an available state for a new proton, the spontaneous decay of a neutron either 'transmutes' the nucleus into that of the element which has one more proton, or causes it to 'fission' into two lighter elements which divide up the electron cloud. In either case, the electron from the neutron decay emerges with so much kinetic energy that it passes through the electron cloud and emerges as beta radiation. Whether or not any nucleus is stable depends, therefore, on whether the Exclusion Principle can accommodate the proton that would result from the decay of one of its neutrons.

COSMIC RAYS

As radiation detectors were refined, it was noticed that there was a 'background' level. It was generally thought that this was radiation leaking from the Earth. In 1911 the Austrian physicist V.F. Hess set out to prove this supposition by sending an electroscope aloft in a balloon to measure the rate at which radioactivity decreased with altitude. To his great surprise, he found that while the intensity of ions in the atmosphere initially declined, it began to increase again above about 1,500 metres. There was evidently "a radiation of a very high penetrating power entering our atmosphere from above". A variety of 'cosmic rays' had been postulated in the nineteenth century, but this was the first convincing evidence for their existence. Nevertheless, it took another decade of research before this was widely accepted.

In 1925, R.A. Millikan at Caltech suggested that cosmic rays were very-short-wavelength electromagnetic waves, but A.H. Compton at the University of Chicago thought them more likely to be very-high-speed charged particles. If electromagnetic, their distribution in the sky would be random, but if electrically charged they would be influenced by the Earth's magnetic field and would tend to concentrate near the magnetic poles. By confirming this 'latitude effect', Compton proved that they are at least partly composed of charged particles. The Italian physicist B.B. Rossi noted in 1930 that if these particles were positively charged they would preferentially arrive from the east. A few months later T.H. Johnson established that this was so. Cosmic rays therefore provided early particle physicists with a natural 'laboratory' for studying high-energy processes.

ANTIMATTER

In the late 1920s, in working out the wave properties of the electron in order to refine Schrödinger's wave mechanics, Dirac found that there were *two solutions*: one concerning the familiar negatively charged electron, and the other involving an electron with a positive charge. Although he initially presumed this other solution to be only a mathematical artefact, in 1930 he realised that it must be real. In the nineteenth century there had been proposals for positively and negatively charged forms of the corpuscles which carried electricity, but this had been mere speculation, and when J.J. Thomson found only negative ones this asymmetry was puzzling. Now it was realised that for some reason the counterpart had eluded detection.

Initially, Dirac's prediction attracted little attention because there was no evidence for such a particle. In 1932, however, C.D. Anderson erected a cloud chamber on a mountain top to be as close as possible to the source of cosmic rays. He also mounted a lead shield in front of the detector to slow the rays sufficiently to enable the magnetic field in the chamber to significantly deflect them. To his great surprise, antielectrons – evidently produced by the cosmic rays interacting with the atoms of lead – emerged from the shield. As a follow-up experiment, a lead sheet was irradiated with alpha particles and electron–antielectron pairs were seen to emerge with their tracks diverging. In 1934 Frédéric and Irène Joliot-Curie discovered that antielectrons were emitted by beta decay of radioactive phosphorus. Although Anderson was a junior member of the Caltech faculty – a student of Millikan – when he made the discovery, the fact that it won him the Nobel Prize in 1936 guaranteed that his application for tenure was favourably received! The symmetry of the electron and antielectron was later generalised to give an 'antiparticle' for each particle, and the introduction of the term 'antimatter' raised the prospect of there being atoms composed of antiparticles.*

A particle and its antiparticle differ in the phase of their wave functions, which are displaced 180 degrees in time. If a particle meets its counterpart, both flash from existence in a process called 'annihilation' in which the interference between the two wave functions reduces both of their amplitudes to zero. Although the mass of the electron is slight, the annihilation of an electron and an antielectron creates sufficient energy to produce a gamma ray. This is because the rest masses of the particles are converted to energy in accordance with $E = mc^2$. The kinetic energy of the collision is comparatively insignificant, but it contributes a fraction of the gamma ray's energy. An argument based on symmetry suggested that a gamma ray of sufficient energy ought to be able to transform spontaneously into a particle and its antiparticle, in a process that was called 'pair production'.

ACCELERATORS

Identifying the antielectron prompted physicists to seek out the antiproton, but while an alpha particle from radioactive decay could smash into an atom of lead with sufficient energy to create an electron–antielectron pair, it was too slow to create a proton–antiproton pair as a proton is 1,836 times as massive as an electron. *Some* alpha particles in cosmic rays were sufficiently energetic, but rather than await a rare event, physicists decided to build 'accelerators' in order to boost the energy of their alpha particles. In this way they became 'farmers' rather than 'hunters'. The first accelerator was constructed in Britain at the Rutherford Laboratory in 1929 by J.D. Cockcroft and E.T.S. Walton. Although it was widely referred to as an 'atom smasher', the term was misleading as its purpose was not to smash atoms but to probe their nuclei. It was, in effect, a factory for alpha particles. Its energised protons could turn lithium-7 into beryllium-8, which split into a pair of alpha particles. In 1930 E.O. Lawrence at the University of California constructed another device,

---

*   Although Anderson proposed the name 'positron', the term 'antielectron' will be used herein to
preclude a proliferation of special names for antiparticles.

named the 'cyclotron', in which an alternating magnetic field bounced protons back and forth along an arc, boosting them on each cycle. The radius of the traverse expanded with each cycle until the protons finally emerged from the periphery to be directed towards a target. In particle physics, the unit of energy is the *electron volt* (eV). The prototype energised protons to 1 MeV (1 million electron volts), but by 1939 this had been increased to 20 MeV. More sophisticated cyclotrons increased the strength of the field in order to hold the protons in the same path as they accelerated. This enabled them to be further energised, and since the strength of the field was synchronised with the acceleration profile these devices were named 'synchrotrons'. Although the prototype built in 1946 at the University of California achieved 400 MeV, in 1952 the proton synchrotron at the Bookhaven National Laboratory on Long Island attained 3 BeV (3 billion electron volts), which corresponded to the mean energy of protons in cosmic rays. In 1954 a similar machine in California attained 6 BeV, thus finally achieving the energy required for proton–antiproton pair production. In 1955 E.G. Segrè and Owen Chamberlain isolated antiprotons by firing protons at a copper target. In the following year Bruce Cork found the antineutron. Synchrotrons were later paired to allow contrarotating beams to intersect, thereby doubling the effective energy. These accelerators were known as 'colliders', and the one that was commissioned in 1982 at the Fermi National Laboratory in Chicago could produce protons of 1 TeV (tera = 10^{12}; or 1,000 billion electron volts). Of course, as the energies increased, so did the pace of discoveries.

THE LITTLE NEUTRAL ONE

Observations in the 1920s had shown that the energy carried off by the electron ejected in beta decay did not account for *all* of the energy released by a nucleus. This violated the law of conservation of energy and was an issue of significant concern. In 1930 Wolfgang Pauli speculated that there was another particle involved in beta decay, and defined the properties that it must have to enable the process to uphold energy conservation: it must be electrically neutral and it must have zero rest mass in order to be capable of carrying arbitrarily small amounts of energy (because *all* of the energy of a *massless* particle is in kinetic form). Furthermore, because each particle in neutron decay was a fermion, it was evident that the spins of two of the products would cancel, leaving a residual $\pm 1/2$ spin to match that of the neutron.

In 1934 Enrico Fermi named Pauli's particle the 'neutrino' (in his native Italian, 'the little neutral one') and refined its properties in the first successful theory of beta decay. Fermi reasoned that just as an atom changes from a higher energy state to one of lower energy by emitting a photon, so a neutron could change into a proton by emitting an electron and a neutrino – in effect, a neutron was a high-energy form of a proton. This idea was so novel that Fermi's paper to the journal *Nature* was rejected by the referee!

In 1929 the American astronomer H.N. Russell spectroscopically determined the composition of the Sun's surface and found that some 90 per cent of the atoms were hydrogen and most of the rest were helium; other elements (what an astronomical

$$n^0 \longrightarrow p^+ + e^- + \overline{\nu}_e^{\,0}$$

The spontaneous decay of a neutron into a proton, an electron and an antineutrino.

spectroscopist considers 'metals') together accounted for less than 1 per cent of the population. Calculations indicated that the temperature in the Sun's core was high enough for fusion processes to turn hydrogen into helium. In 1935, H.A. Bethe, who had been a student of Sommerfeld, fled Germany to America and accepted a post at Cornell University. In 1938 he formulated the details of how this process worked. In fact, *several hundred tonnes* of matter is converted into energy every second. Although neutrinos stream from the Sun in enormous numbers, no one had ever noticed one because they do not interact electromagnetically: one would be able to fly through a block of lead billions of kilometres thick without interacting. Tracking down something as elusive as this would obviously pose a major technical challenge. However, their existence could be inferred indirectly. When a neutrino turned a proton into a neutron it would release an antielectron.

$$p^+ + \overline{\nu}_e^{\,0} \longrightarrow n^0 + e^+$$

An antineutrino can turn a proton into a neutron and an antielectron.

In 1953 Frederick Reines, C.L. Cowan and Maurice Goldhaber employed a tank of water (effectively a sea of protons) at Los Alamos as a neutrino detector. They instrumented it to record the gamma rays from the annihilation of the antielectrons from the neutrino interaction by ambient electrons, and put it near a reactor. In 1956 they confirmed the existence of the neutrino. It was a *tour de force* case of the power of accepting the universality of a principle and working through the consequences.

4

Nuclear forces

EXCHANGE PARTICLES

By 1932 quantum mechanics had explained atomic structure in exquisite detail, and the subatomic realm had been found to consist of electrons, protons and neutrons plus the postulated neutrino as fermions, and the photon of electromagnetism and the hypothetical graviton of gravitation as bosons, with what appeared to be a pattern of two lightweight fermions (electron and neutrino), two heavy fermions (proton and neutron) and two bosons. Following a quarter of a century of shocking discoveries, it began to seem as if a *complete theory* of the subatomic realm was about to be finalised. The outstanding issue was the force that bound the neutrons and protons in the nucleus. As electromagnetism is inversely proportional to the square of the distance between two charges, the electrostatic repulsive force between two protons is exceedingly potent at a separation of 10^{-15} metre. Early calculations indicated that the attraction of this 'strong nuclear force' (subsequently abbreviated to the 'strong force') was about 100 times stronger than the electrostatic repulsion. In 1928 P.A.M. Dirac had described electromagnetism in terms of charged particles exchanging photons. Following the verification of the neutron's existence in 1932, W.K. Heisenberg speculated that the protons and neutrons in the nucleus might exchange some type of particle and thereby flip their identities back and forth so rapidly that the electrostatic repulsion between the protons was ineffective, but calculations showed that this was impractical. Nevertheless, the 'exchange particle' idea was intriguing, and in 1934 Hideki Yukawa, a theoretical physicist at Osaka's Imperial University in Japan, proposed this as the mechanism of all quantum forces. Einstein had established that energy and mass were interchangeable by $E = mc^2$, and if Heisenberg's Uncertainty Principle was reformulated in terms of energy as $\Delta E \Delta T > h$, then it was evident that instead of being starkly empty, the vacuum must be alive with 'virtual particles' which flit in and out of existence. A particle–antiparticle pair with a total energy ΔE can spontaneously be manifested in the vacuum and they can roam freely, but they must annihilate within the corresponding

ΔT interval. Because nature is indifferent to such fleeting particles as long as they do not interact with 'real' particles, such shenanigans do not violate the conservation of energy. The maximum range that a virtual particle could travel at the speed of light, c, is $c.\Delta T$, and because more massive virtual particles must borrow a larger quantity of energy from the vacuum, the time-scale is brief and the distance to which they can roam is more limited. Yukawa posited that the electromagnetic force arises because electrically charged particles constantly exchange virtual photons, and the number of exchanges increases as the range between the particles decreases. The effect of this interaction is attractive for opposite charges and repulsive for matching charges. He realised that the strength of the strong force must decline more rapidly than with the square of the range – in fact, doubling the range reduces it by a factor of 100.

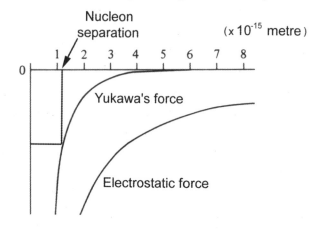

The force that Hideki Yukawa realised held the nucleus together was much stronger at short range than the electrostatic repulsion that would otherwise split it apart.

While this force is extremely potent within the nucleus, its strength declines so steeply that its effective range is 1/100,000th of the diameter of the atom which – by no coincidence – is comparable to the size of the nucleus. It is the great strength of this attraction at short range that makes the nucleus such a *compact* structure, with the protons and neutrons jostling one another.

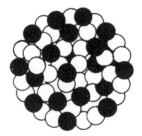

Uranium is one of the largest nuclei.

Yukawa also realised that while the infinite range of the electromagnetic force required a massless exchange particle, the short range of the strong force implied that its exchange particle must possess mass. His calculations indicated a mass several hundred times that of an electron, or about 15 per cent of the mass of a proton. Because tests had shown that the strength of the force was the *same* between a pair of protons as it was between a pair of neutrons – or indeed between a proton and a neutron – it was evident that there were both positively and negatively charged exchange particles. If such a particle and its antiparticle formed in the vacuum of the densely packed nucleus, and if one were to be absorbed by a proton and the other by a neutron, then it would *appear* to a macroscopic observer that the proton and neutron had switched identities by exchanging a single particle.

Further, while the electromagnetic force repels like-charges, the strong force is *always attractive*. This is a characteristic of spin $= 0$ and spin $= 2$ bosons but not of spin $= 1$ bosons, and explains why the strong and gravitational forces are universally attractive while the electromagnetic force depends on the polarity of the electric charges. In addition to diluting the electrostatic repulsion by interposing themselves between the protons, the neutrons actively contribute to the attractive force that binds the nucleus together. If this was correct, the protons and neutrons would be *immersed within a cloud of virtual bosons*. However, because no intermediate-mass particles had ever been detected, Yukawa doubted the validity of his theory and his paper went unnoticed not only in America and Europe but also in Japan.

MUON

Cosmic rays include particles that stream out from the Sun and particles from sources beyond the Solar System. The 'solar wind' plasma comprises mainly electrons and protons. The external material is primarily protons, but they are more energetic as a result of having been accelerated to relativistic speed by the magnetic fields in space, and when one penetrates the Earth's atmosphere it undergoes a succession of encounters with atomic nuclei, progressively shedding its energy as a spray of particles, some of which are sufficiently energetic to do likewise in a 'cascade'. A particularly energetic cosmic ray can generate *millions* of products, whose masses, kinetic energies and the recoils of the atoms with which they interact all derive from the kinetic energy of the original particle.

A major cascade can reach sea level, but can more readily be investigated from a mountain top. In 1934 C.D. Anderson, who was still studying cosmic rays after finding the antielectron, noted 'anomalous' tracks in a cloud chamber that indicated the passage of a negatively charged particle whose path curved less than that of an electron, implying that it was significantly more massive than the electron. Anderson's student, S.H. Neddermeyer, then modified the apparatus to slow the particles prior to entering the chamber so that their tracks could be more readily deflected. In 1936 they announced the discovery of a particle with a mass some 207 times greater than the electron. As this was 11 per cent of the proton's mass, it was named the 'mesotron' (the 'meso' prefix being Greek for 'intermediate'). In 1939,

The enormous energy of a primary cosmic ray is dissipated as it passes through a stack of brass plates in a cloud chamber in a cascade of secondaries.

however, the Indian physicist H.J. Bhaba pointed out that a more linguistically accurate name was 'meson'.

In January 1937 Yukawa, correctly suspecting that Anderson was unaware of his paper, suggested that the new particle might be the exchange particle of the strong force, and for a while this was widely presumed to be the case. However, an experiment that was performed in the basement of the Vatican in Rome in 1945 by M. Conversi, E. Pancini and O. Piccioni showed that it did not strongly interact. In fact, Heisenberg and Hans Euler had noticed in 1938 that it would travel through a target for up to 10^{-6} second without interacting with a nucleus before spontaneously decaying into an electron and a pair of neutrinos. This was a remarkably long life for a particle. In 1947 Enrico Fermi, Edward Teller and V.F. Weisskopf calculated that the process by which it decayed was *10^{13} times weaker* than the strong force, for which the characteristic time was 10^{-23} second. Shoichi Sakata and Takesi Inoue in Japan and H.A. Bethe and R.E. Marshak in America independently concluded that this could not possibly be Yukawa's particle. In 1953, when it was recognised to be a 'heavy electron' it was renamed the 'muon'. "Who ordered that?" asked I.I. Rabi of Columbia University in New York, expressing the frustration that it appeared to play no role in the structure of matter.

PION

In 1947, following Anderson's example of taking a cloud chamber up a mountain to study cosmic rays nearer their source, C.F. Powell of the University of Bristol in England and a former student of Rutherford, sent a newly manufactured sensitive film emulsion to be exposed at the astronomical observatory on the 3,000-metre summit of Pic du Midi in the French Pyrenees. In addition to the expected muons, he found a particle with 273 times the electron's mass. There were two types, one positively charged and one negatively charged, and experiments confirmed that it *did* interact with nuclei. Although this 'pi-meson' (abbreviated to 'pion') was responsible for the force that bound the nucleus together, if one became isolated it would spontaneously decay into a muon after 10^{-8} second.

Yukawa's particle had taken so long to find because (1) it could not travel far in a dense medium without strongly interacting and (2) it could be detected in a cosmic ray cascade only near its point of origin – and hence was best sought from a high-altitude site. However, the most advanced cloud chambers were bulky structures unsuitable for field stations. Furthermore, because the gas in a cloud chamber had to be at a low pressure, a pion was *less* likely to encounter a nucleus than in the film emulsion, and pions travelling through a cloud chamber left tracks that were difficult to distinguish from those of muons. In any case, high-energy physicists had been preoccupied for much of the decade with the Second World War in general, and with the development of the atomic bomb in particular.

W.K. Heisenberg had pointed out in 1932 that if electric charge was ignored, the proton and neutron could reasonably be considered to be two 'states' of a single *nucleon* particle. In 1937 the Hungarian physicist E.P. Wigner at Princeton proposed that they were analogous to *isotopes* and their distinction could be defined in terms of an abstract *isotopic spin*. In 1938 Nicholas Kemmer in Russia noted that if the charged versions of Yukawa's exchange particle were thought of as two states of isotopic spin, then there ought to be a third, electrically neutral, version. In 1949 R. Bjorkland used Berkeley's synchrotron to manufacture pions. A neutral pion was expected to decay into gamma rays within in 10^{-16} second. He emplaced a metal foil in a position to induce these gamma rays to create electron–antielectron pairs, which they did. In 1948 J.R. Oppenheimer suggested that cosmic gamma rays were largely the result of the decay of neutral pions in cosmic rays.

Although the recognition of the strong force increased the number of fundamental forces to three, by 1950 physicists were feeling rather pleased with themselves and it once again began to look as if all that remained to be done was to refine the details of their theories. However, they were about to receive a nasty shock.

STRANGE PARTICLES

In 1944 Louis Leprince-Ringuet and Michel l'Héritier in France reported a single cosmic ray track suggestive of a 'heavy' meson, but this prompted little interest. In studying cosmic rays in a cloud chamber in 1946 and 1947, G.D. Rochester and C.C. Butler in Manchester noticed tracks of a particle with a mass of about 800 times that of the electron, which was almost half the mass of a proton; they dubbed it the 'V' particle. In 1948 Powell identified on film the track of a particle with a similar mass which decayed into a trio of pions; he named it the 'tau-meson'. The existence of a heavy meson was confirmed by Anderson in 1950. By 1953, it was clear that these were all related, and they were named the 'K-mesons' (abbreviated to 'kaons').

Meanwhile, in 1951 Butler had identified the neutral lambda particle, which was about 20 per cent more massive than the proton. As it decayed into a neutron and a neutral pion, it was initially regarded as a 'heavy neutron'. The negatively charged xi particle found in 1952 was even heavier. Although these heavy particles were first discovered in cosmic rays, the introduction of the Cosmotron at Brookhaven in 1953 facilitated a systematic search – with a profusion of discoveries. The positive sigma (intermediate in mass between the lambda and the xi) was found in 1953; its negative partner was identified a few months later; and the neutral one was added in 1956. The neutral xi was finally tracked down in 1959, but efforts to find a positive version failed. In 1953 these heavy versions of the nucleons, all of which were unstable, were collectively named 'hyperons', but what role did they play?

Although the kaons and hyperons were created in strong interactions and hence *ought* to have been able to interact in this way, they rarely did so; they drifted until finally decaying after 10^{-10} second by a process that was *10^{13} times weaker* than the strong force – a strength that, intriguingly, was the same as the process by which the muon decayed. The reason for this difference between the processes of production

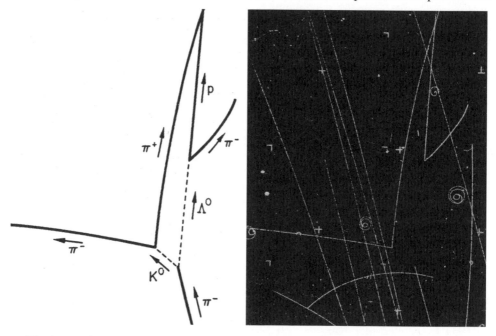

The decay of two particles by the weak interaction in a bubble chamber picture by L.W. Alvarez at the University of California. An energetic negative pion from the Bevatron strikes a proton, giving rise to a neutral kaon and a neutral lambda which leave no tracks. The kaon decays into a negative pion and a positive pion. The lambda decays into a proton and a negative pion. Both are weak decays. Courtesy of the Lawrence Berkeley Laboratory.

and decay was a mystery. In 1952 Yoichiro Nambu and Kazuhiko Nishijima in Japan and Abraham Pais in America independently noted that while the strong force could create or consume such particles *in pairs*, once they were isolated they could not interact in this way, they drifted and spontaneously decayed. This 'associated production' phenomenon was confirmed by experiments at Brookhaven in 1953, and for the first time physicists realised that there were *two* nuclear forces at work, with the weaker one (promptly dubbed the 'weak force') apparently being related to beta decay. The isolated long-lived 'strange' particles seemed to be in a 'metastable' state; something seemed to be *inhibiting* strong interactions. What? A child prodigy and polymath, Murray Gell-Mann, was accepted by Yale at the age of fifteen. After gaining his doctorate from the Massachusetts Institute of Technology in 1951 (at age 21) he joined the faculty at the University of Chicago. In 1953 he and Kazuhiko Nishijima and Tadao Nakone in Japan independently proposed another quantum number, which Gell-Mann named 'strangeness'. In effect, 'normal' mesons and nucleons had a strangeness = 0. Although strange particles could be created by the strong interaction, the fact that it was a conserved quantity meant that it could be created only in equal and opposite amounts, which was why strange particles were produced in pairs. Likewise, such particles could be consumed in strong interactions only if there was no overall change in strangeness. The lengthy lifetimes were a consequence of the strength of the weak force. Significantly, while the strong force conserved strangeness, the weak force enabled it to 'leak away'. After many years of toil, the strong force seemed to be explained by Yukawa's meson, but how this new force operated was not known.

QUANTUM ELECTRODYNAMICS

P.A.M. Dirac's 1928 theory of the hydrogen atom was generally satisfactory in terms of accounting for Sommerfeld's fine structure formula, but it did not take into account the fact that the electron would *interact with its own field*, and the equations were plagued by infinities that rendered its results meaningless. In the 1930s a significant departure from the theory was noted in the H-alpha line. W.E. Lamb of the Columbia Radiation Laboratory found that two lines that should have precisely the same wavelength, were actually slightly separated. This became known as the 'Lamb shift'. When he presented his observations at a conference in June 1947 theorists in the audience realised that this might be due to the electron's interaction with its own field by constantly emitting and absorbing virtual photons. H.A. Bethe was first to verify that this was indeed so, but his calculations were not relativistic and in any case suffered from infinities. Julian Schwinger at Harvard measured the electron's anomalous magnetic moment, which a group at Columbia showed was also at odds with Dirac's theory. Schwinger developed a consistent theory that was free of infinities and predicted both the magnetic moment and Lamb's shift, and he presented it to a conference in 1948. S.-I. Tomonaga in Japan independently derived essentially the same formulation. After being awarded a doctorate from Princeton in 1942 under the supervision of J.A. Wheeler, and then spending the following few years at Los Alamos working on the atom bomb, R.P. Feynman moved to Cornell, where he developed a very different approach to a theory, using a diagrammatic tool (dubbed the 'Feynman diagram') to represent particle interactions. His procedure was to *sum* the probabilities of all the ways in which an interaction could occur, each way being represented by a different diagram. He found that every positive infinity in the calculation was cancelled out by a corresponding negative infinity, yielding a sensible solution. In 1949 Freeman Dyson at Cornell showed the Schwinger–Tomonaga and Feynman formulations to be equivalent, and in the process he expressed *quantum electrodynamics* in terms of fields. The result was an enhancement rather than a rejection of Dirac's theory: where the original was computable they both gave the same result, with the revision producing greater precision; and where the original was rendered non-computable by infinities, the revised – renormalisable – theory gave sensible results. As Dyson reflected, quantum electrodynamics "gives us a complete description of what an electron does [and] therefore in a certain sense it gives us an understanding of what an electron is". In fact, the precision was so great that he felt that it gave "some grasp of the nature of an elementary particle". The success of quantum electrodynamics prompted the expectation that quantum field theory could be applied to account for the other fundamental forces, but this optimism soon waned.

5

Symmetries and phase changes

A NEW OUTLOOK

If a *transformation* of a system yields no observable change, it is said to exhibit a *symmetry*. Group theory rigorously describes symmetries mathematically, but using a rather arcane language. In 1872, while writing his thesis *On a Class of Geometric Transformations*, the Norwegian mathematician M.S. Lie noted that symmetries in differential equations could be described by a continuous group of transformations – a construct called a Lie group. In 1894 E.J. Cartan in France generalised Lie groups, and methodically classified them from simple mirror-image symmetries to complex rotational symmetries. Emmy Noether of Göttingen University pointed out in 1915 that where a symmetry occurs in nature, some associated property is conserved. In fact, this can serve as a definition: *a symmetry is a property of a system that remains invariant with respect to a transformation of some kind.*

The symmetry of a snowflake is such that its appearance is unchanged when it is rotated through 60 degrees, and six such transformations are equivalent to a complete 360-degree rotation. This symmetry derives from the angle between the hydrogen atoms in the H_2O molecule. Prior to their forming a water molecule, the distributions of electric charge *within* the individual atoms are rotationally

Three of the complex crystalline structures that snowflakes can form.

symmetric, but become polarised as they come together, and the breaking of this symmetry manifests a force that re-establishes the balance by locking the atoms into an asymmetric composite structure. A water molecule is electrically neutral but as its charge distribution is *not* uniform, this creates an electrostatic force which enables neighbouring molecules to form bonds (in the case of liquid water, these are called van der Waals bonds). By constraining the motions of the molecules, these loose bonds give water the fluidic properties of a liquid. If the water is heated to its boiling point the random motions of the molecules become sufficiently vigorous to break these bonds and the liberated molecules act as a gas. Similarly, if water is cooled to its freezing point the molecules will become so sluggish that they link up to establish crystals which immobilise the atoms in a regular structure. These three states of matter – gas, liquid and solid – are known as *phases*, and the transitions are *phase changes*. Freezing is a *spontaneously* broken symmetry because the orientation of the molecules that start the process of crystallisation randomly selects the orientation of the resulting structure. In 1760 the Scottish chemist Joseph Black realised that when water freezes there is more energy in the collection of molecules of liquid water at 0 °C than in the crystalline structure of ice at the same temperature. This energy is called the *latent heat* of crystallisation. It is liberated during crystallisation, and must be imparted to melt ice. Melting restores the symmetry lost in forming a crystal. Latent heat is also released in condensation, and it must be reintroduced during vaporisation at 100 °C to reinstate the lost symmetry. If water vapour is heated, its molecules will dissociate into their constituent atoms, and continued heating will yield a plasma of ions and free electrons. The properties of matter are therefore temperature dependent. A phase change in which latent heat is required to reorganise the structure of a material is referred to as a *first-order* phase change, and the physicial properties of the material change *discontinuously*.

A *second-order* phase change, in which there is no physical restructuring and no latent heat, proceeds more *smoothly*. An atom is a tiny magnet because its electrons generate a magnetic moment. This effect is particularly strong in the case of iron. In 1895 Pierre Curie discovered that an iron magnet loses its magnetism upon being heated above 765 °C and regains it upon cooling. As the melting point of iron is 1540 °C, this phase change occurs *in* the solid state. The individual atoms of hot iron are oriented randomly and so their magnetic moments tend to cancel out and there is no macroscopic magnetic field. As the kinetic energy of the atoms declines as the metal cools, however, the individual motions of the atoms slow sufficiently for the weak magnetic force between them to dominate. The atoms do not adopt another physical structure, they remain free to move, but by adopting a specific orientation they yield up a component of their angular momentum. As the strength of the macroscopic field created by mutually aligning the individual magnetic moments grows *gradually* as the temperature approaches the critical value, this is a smooth transition. When an electrically charged particle is accelerated by a magnetic field, some of the field's energy is converted into the kinetic energy of the particle, and the field is diminished accordingly. Conversely, the rotational energy shed as the iron atoms adopt the same orientation is *transferred* into the macroscopic magnetic field. There is no *net* change in energy in cooling through the Curie point, but the mutual

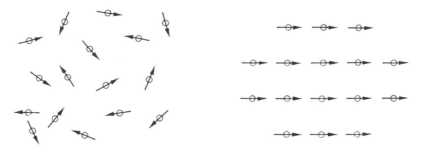

In hot iron (left) the magnetic axes of the individual atoms are randomly aligned and there is no overall magnetic field, but as it cools (right) they become mutually aligned and a macroscopic field become manifest.

realignment of the atoms involves a *redistribution of energy between atoms and the magnetic field.*

A *hot* iron bar could *be* magnetised, but this would require the application of an *external* magnetic field to force the atoms into a single orientation, and the symmetry breaking would not be spontaneous because the induced field would line up with the field that supplied the energy. In terms of magnetisation, therefore, hot iron is in its *lowest energy state* because the randomly aligned magnetic moments cancel and the overall energy of the magnetic field is zero. Magnetisation can be induced in hot iron only by *adding energy* to the system, so the energy-versus-magnetisation diagram is a U-shape in which the lowest energy state corresponds to zero magnetisation.

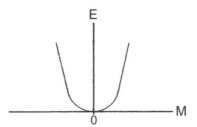

In hot iron, the minimum energy state is at zero magnetisation.

In the phase change that occurs on cooling through the Curie point, the diagram adopts a profile in which the zero of magnetisation occurs at a positive energy state, and the lowest energy state has a *non-zero* magnetisation (there are in fact two situations, corresponding to whether the field is aligned 'up' or 'down'). The phase change is marked by the transition from one relationship to another.

Systems tend towards their minimum state for *all* forms of energy. The *ground state* for 'hot' iron has zero magnetisation at the minimum of energy, but there is a non-zero magnetisation for 'cold' iron. The energy that was transferred into the field during the phase change *cannot be shed by further cooling*, and this '*frozen*' field is therefore a '*false*' ground state. When random motions become sufficiently vigorous

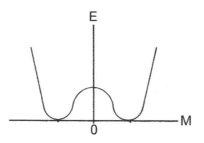

When iron cools through its Curie temperature, the lowest energy states occur at non-zero degrees of magnetisation.

to break the magnetic bonds upon being reheated, the spatial symmetry is restored and the strength of the macroscopic field will gradually fade to zero.

The symmetries expressed by group theory can be more general (or 'abstract') than geometrical structures and their rotations in space. When a pion interacts with a nucleon, it is indifferent to whether this is a proton or a neutron. In terms of group theory, nucleons can be represented by a single group. If we switched our definitions of protons and neutrons, it would make no difference to the pion. The property that distinguishes between nucleons is isotopic spin. Switching could be considered as the rotation of a 'selector arrow' in an abstract space. From our point of view, isotopic spin can be thought of as a broken symmetry, but to the pion this symmetry is *exact*. The strong force is therefore said to be invariant under rotations in isotopic spin space.*

GAUGE INVARIANCE

The concept of 'gauge' (or scale) invariance was introduced in 1918 by H.K.H. Weyl, in an unsuccessful attempt to unify electrodynamics with general relativity. In 1929, after the development of quantum mechanics, he reformulated the concept of invariance to mean that *the choice of phase of a wave function does not affect the wave equation*. This, however, was largely ignored until 1954, when C.N. Yang and Robert Mills, young theorists at the Institute for Advanced Study at Princeton who were inspired by the newly completed theory of quantum electrodynamics, developed a *gauge theory* that was invariant with respect to transformations depending upon position and time and, hence, was insensitive to the manner in which its properties were measured – a requirement which they expressed as the Principle of Gauge Invariance. Yang and Mills realised that a force field is how nature maintains some associated symmetry *at each and every point in space*. A force is simply nature's way

* More precisely, the strong force is invariant under SU(2), where 'SU' indicates the 'special unitary' group ('unitary' being one of Cartan's three classes) and the '2' indicates that there are two degrees of freedom.

of expressing a global symmetry in a local situation. The symmetries corresponding to Cartan's groups can *all* be expressed as Yang–Mills fields. As noted, there are several kinds of field. A scalar field has a value at every point in space, but no intrinsic directionality. A particle in a uniform scalar field will not feel a force, but if there is a *gradient* in the field this broken symmetry manifests itself as a force which seeks to restore uniformity. Temperature is a scalar field. If there is a gradient in a temperature field, the resulting force will cause material to move from warm to cool regions. Similarly, in mechanics, stress is the force induced by the gradient in a strain field. On the other hand, a magnetic field has a vector and so an electrically charged particle in a uniform field will be accelerated along the 'line of force'.

The choice of Greenwich in England for the 'prime meridian' was arbitrary but, once it was defined, longitude became a *global gauge symmetry*. It would be *possible* for each country to define the prime meridian as passing through its own capital, but complications would arise at their borders. Such complications *do* arise as a result of different countries having different legal systems. For such localisation not to matter, there would have to be a comprehensive means of mapping laws of one system into another. Lacking such a comprehensive transformation, human efforts at establishing a global legal system exibiting *local gauge symmetry* are fraught with conflicts.

A global gauge symmetry can have only one 'zero'. Although each observer can define this arbitrarily within a given region of space (as a nation can claim the prime meridian) an electrical *force* would arise at the border between regions in which the zero of potential was defined differently. But such a force does *not* arise. Although electricity does not exhibit local gauge symmetry, electricity is merely one aspect of electromagnetism; the other being magnetism. Changing the definitions pertaining to one changes those pertaining to the other and (it turns out) the force that would be expected to arise from differing local definitions of electric potential is mathematically *cancelled* by the corresponding changes in the magnetic potential. Neither electricity nor magnetism exhibits local gauge symmetry but electromagnetism does. In terms of Maxwell's equations, changes in the electric potential arising from arbitrary changes of 'zero' at each point in space are precisely cancelled by changes in the magnetic potential, and so we are free to define zero as we chose, it is of no importance.* Since electromagnetism has a *higher* symmetry it can be thought of as being *simpler* than either of its physical manifestations; electromagnetism is a 'hidden' symmetry. Lest this idea of cancellations bestowing the freedom to set the zero of electrical potential differently at different locations appear dubious, consider special relativity in which Einstein noted that space and time are related and measurements of one influence the other. Specifically, observers moving with respect to one another at a constant speed have different frames of reference. At relative speeds approaching the speed of light, space and time become malleable so that even though observers

---

* As a result of local gauge symmetry, electromagnetism is invariant under the unitary group U(1), a one-dimensional abstract space in which the two directions correspond to whether electric charge is positive or negative.

in different frames of reference may disagree as to *specific* lengths and time intervals, *within their defined scales* they will agree on the invariant of a local gauge symmetry – the speed of light. In retrospect, the Principle of Relativity is a direct consequence of the gauge invariance of the electromagnetic field.

Because Maxwell's equations do not encompass quantum mechanics, they form 'classical' field theory. Yang and Mills argued that as classical field theory exhibited a local gauge symmetry, this must also be exhibited by quantum field theory. Since force is a *manifestation* of a symmetry, they realised that a local gauge symmetry *required* a Yukawa-style mediator particle. The phenomenon of electromagnetism with its photon is therefore a consequence of the existence of a field that we perceive as electric charge.

Suspecting the generalisation that all fundamental forces are manifestations of symmetries, Yang and Mills reasoned that for every conserved property (such as the quantum number of a fundamental charge) there must be an associated force that acts to maintain the invariance. In this case, each fundamental force must be related to a symmetry in some space, either physical or abstract. Yang and Mills tried to develop a gauge theory for the strong force as an extension of electrodynamics. Although their theory predicted a trio of exchange particles, in contrast to Yukawa's spin = 0 pions, these spin = 1 bosons were massless, which meant that, just like the photon, they would have infinite range. In this respect, the theory failed to account for the force's extremely short range.

THE WEAK FORCE

In 1933 Enrico Fermi suggested that beta radioactivity, and the manner in which a neutron spontaneously decayed, could be described using a formalism similar to that developed by Dirac for the electromagnetic force, but 10^{-10} weaker. With its range of only about 1/1,000th the diameter of the nucleus, it could not play a role in binding the nucleus, but it could affect *individual nucleons*. The fact that the metastable particles exhibited the same characteristic time of 10^{-10} second indicated that this weak force acted on *many types* of particle.

In addition to transforming a neutron into a proton and vice versa, the weak force was evidently responsible for the decay of a muon into an electron. As *all* of these interactions involved neutrinos (which were unique in interacting *solely* by the weak force) it was presumed that the neutrino was characteristic of the weak interaction, but when the strange particles were found to decay without neutrinos it was realised that the force was more complex than had been posited in Fermi's theory of beta decay. Nevertheless, the fact that aspects of the force could be explained by an electromagnetic formalism implied that these forces shared an *underlying symmetry*. The form of this symmetry was obscure, but so too had been the relationship between electricity and magnetism until Maxwell recognised it. In 1938 Oskar Klein had mused that beta decay might involve a spin = 1 boson. If such a particle was a counterpart to the photon, then the underlying symmetry might be a gauge theory. In 1957 Julian Schwinger at Harvard suggested that if the weak force had a pair of

Table 5.1 Relative strengths of forces

Field	Relative strength	Characteristic time (second)
Strong	1	10^{-23}
Electromagnetic	10^{-3}	10^{-20}
Weak	10^{-13}	10^{-10}

Notes:
1 The range of the strong force (1 fm) corresponds to a 'characteristic time' of about 10^{-23} second, this being the time for an interaction across a nucleus 3 fm in diameter; an 'event' taking place in a shorter time has 'no meaning'.
2 For electromagnetic interactions, the strength is 10^{-3} of the strong force, and so the characteristic time is longer (10^{-20}); this is roughly the time for a photon to cross an atom.
3 Weak interactions have 10^{-13} of the strong force, and the characteristic time is 10^{-10} second, which is a veritable eternity in the realm of particle interactions. The range of the weak force is very short, only slightly greater than the diameter of a nucleon.

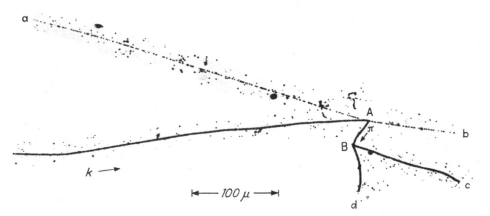

In 1949 C.F. Powell's research team spotted this rare example of kaon decay. The ionisation induced by the passage of the energetic charged particles is revealed by a succession of illuminated grains in the photographic emulsion. There is a background of grains, but the tracks stand out well. Their lengths and geometry were measured using a microscope. The kaon (k) enters from the left and decays into a trio of pions at point A, two of which fly off at high speed in opposite directions, leaving faint tracks, while the third flies off more slowly, leaving a denser track, and prompts nuclear disintegration at point B.

electrically charged exchange particles (which he named $W^{\pm}$) then by regarding these as forming a triplet with the photon it might be feasible to link it to electrodynamics in a unified gauge theory. However, he faced the same problem as had stymied Yang and Mills in their effort to describe the strong force, in that the theory's massless vector bosons were incompatible with a short-range force.

The 'parity' of a particle is an abstract property, but it can be regarded as being analogous to mirror-reflection, with 'left-handed' and 'right-handed' values. In 1927

E.P. Wigner posited that parity was a conserved quantity in interactions, but in the early 1950s some kaons were observed to decay in a manner that violated parity conservation. Parity was conserved when a kaon decayed into a pair of pions, but in a few rare cases when it decayed into a trio of pions parity was *not* conserved. In 1956 Yang and T.-D. Lee, a fellow theorist at Princeton, realised that the weak force need *not* conserve parity. In 1957 Chien Shung Wu and L.M. Lederman at Columbia investigated parity in the case of cobalt-60 beta decay. They first applied a magnetic field over the sample in order to orient all the atoms in the same direction, the sample was then chilled to within a whisker of absolute zero to damp out thermal motions and the ejected electrons were found to have a preferred 'handness'. To determine which property *was* conserved by the weak force, it was posited that if a particle had left-handed parity (P) then its oppositely charged (C) antiparticle would be right-handed, and that would conserve CP. However, in 1964 V.L. Fitch and J.W. Cronin at Princeton observed 'forbidden' decay modes of kaons that indicated that the CP symmetry was also violated. Nevertheless, it was realised that if 'time's arrow' (T) was included, then CPT *was* conserved. This implied a gauge invariance in which the changes caused by breaking C were cancelled out by the changes resulting from breaking P, and CPT was conserved by breaking T to cancel the effects of violating CP. The component symmetries were rather more abstract than electricity and magnetism in the case of electromagnetism, but the *same principle* was at work. In 1957, Murray Gell-Mann and R.P. Feynman, both now at Caltech, described the weak force in terms of how it depended on the orientations of the spins of the interacting particles.* A similar idea was independently devised by R.E. Marshak and E.C.G. Sudarshan at the University of Rochester. The violation of parity was accounted for by obliging the neutrino to be intrinsically left-handed. In addition to explaining all *known* forms of the force, the theory had something else going for it, in that several other interactions, which were possible in principle but were inconsistent with vector bosons, were *not* observed. In 1959 the theory received support from an experiment: it had been predicted that one in every 10,000 charged pions would decay into an electron rather than into a muon, and this was confirmed by a team led by the Italian physicist Georgio Fidecaro using CERN's powerful Proton Synchrotron. Despite its success, the theory was inelegant and its computational utility was limited because it resisted efforts to make it renormalisable. It was evident that something was missing, but what?

In September 1958 S.L. Glashow, a student of Schwinger, submitted his doctoral thesis on *The Vector Bosons in Elementary Particle Decay*, and set off with a two-year fellowship from the National Science Foundation for the Neils Bohr Institute in Copenhagen. After graduating from Lahore, Pakistan, Abdus Salam received his doctorate in mathematics and physics from Cambridge in 1952 and then became professor of theoretical physics at the Imperial College of Science and Technology in London. In 1959 Glashow met Salam, and explained why he thought a gauge theory that unified electrodynamics and the weak force *should* be

---

* This is also known as the V–A theory (which is pronounced 'V minus A'), indicating the labels of the orientations involved.

renormalisable, but Salam was not impressed. In early 1960, Glashow met Gell-Mann at the Collège de France in Paris. After a discussion of the issue of renormalisation, Gell-Mann invited him to attend Caltech on a research fellowship after his work was finished in Copenhagen. A few months later, Glashow wrote a paper in which he discussed the prospect of unification. Putting aside the issue of renormalisation, he focused on the underlying symmetry. While at first sight there seemed to be little or no similarity between the phenomena associated with the weak interaction and electromagnetism, if both were mediated by vector bosons some intriguing parallels emerged: both interactions were universal and a single 'coupling constant' could describe a wide class of phenomena. In effect, if the weak force was mediated by electrically charged bosons, the weak interactions would constitute an electric current. After crediting Schwinger for proposing a triplet of spin$\,=1$ vector bosons for both the weak and electromagnetic forces, he pointed out that certain recent experimental results could be explained if there was "at least one neutral intermediary". Glashow suspected that the 'neutral current' was likely responsible for the weak interactions violating parity, while electromagnetic interactions conserved it. However, there was no direct evidence for such a neutral current and the paper drew little interest. In fact, as there were already an astounding number of particles, no one was inclined to invent hypothetical ones.

Despite the success of quantum electrodynamics, physicists had by the mid-1950s become disenchanted with quantum field theory, partly because the 'old hands' such as Dirac believed that even though proper results were predicted by eliminating the infinities by renormalisation, this was a mathematical trick. In addition, efforts to apply quantum field theory to the strong force had met with little success, and the version for the weak force was not renormalisable. By 1960, therefore, quantum field theory was out of favour, and in 1965 Freeman Dyson, now disillusioned, predicted "in a few years the concepts of field theory will drop totally out of the vocabulary of day-to-day work in high-energy physics".

THE HADRON ZOO

In a paper in 1949 entitled 'Are Mesons Elementary Particles?', Enrico Fermi and C.N. Yang had perceived a looming crisis. In 1952 D.A. Glaser at the University of Michigan invented the 'bubble chamber'. In essence this was the inverse of the cloud chamber. Instead of a charged particle's passage condensing droplets of liquid from the gaseous phase, a tank of supercritical liquid hydrogen was heated almost to its boiling point so that the energy imparted by the passage of a particle left a trail of bubbles to be recorded by flash photography. (He is reputed to have been inspired by the manner in which trails of bubbles rise in a glass of beer.) Experiments by L.W. Alvarez at Berkeley turned up such a profusion of new particles that in 1962 L.B. Okun in Russia introduced the term 'hadron' for particles which were susceptible to the strong force.

Why were there so many? If the nucleus consisted of only protons, neutrons and pions, what role did all the others play? Surely they could not all be 'elementary'?

Were some more elementary than others? Was there a subatomic equivalent to the Periodic Table of Elements?

In 1943 W.K. Heisenberg set out to free his matrix mechanics theory of infinities in order that it could serve as a rival for relativistic quantum electrodynamics. Unlike quantum field theory – which relied upon *un*observable concepts such as local field operators, continuous space-time and causality – S-matrix ('S' for scattering) theory employed only *observable* quantities. When quantum electrodynamics was rendered renormalisable, the theory was dismissed as unnecessarily cumbersome, but in 1961, by which time quantum field theory was floundering, G.F. Chew at Berkeley revived it as 'analytic S-matrix theory'. Over the next five years, he developed it to describe the hadrons. As Chew *ignored* the space-time continuum in his formulation, he did not use the wave function whose phase is the essence of quantum field theory. His objective was "to understand nature not in terms of fundamentals but through self-consistency". He argued that a *consistent* theory that matched observable quantities would naturally specify the masses, spins and electric charges of the hadrons in their variety. In fact, Chew rejected the traditional reductionistic view that some particles were the 'building blocks' from which all others were constructed as aggregates. He argued that they were *all equally elementary*. This line of thought was prompted by the impracticability of hierarchically ordering the proliferating 'zoo' of hadrons. As a result of considering each particle to 'potentially contain' any combination of others, it became known as the 'bootstrap theory'. However, despite its ability to explain hadronic interactions, the theory had nothing whatever to say about leptons and the weak and electromagnetic forces, and it was fundamentally incompatible with quantum electrodynamics. Fortunately, when presented with puzzling observations, theorists are able to invent a multiplicity of theories, whose survival depends upon how well they explain what is known and provide new insights.

After striving in vain to develop a gauge theory to unify electrodynamics and the weak force, in late 1960 Glashow went to Caltech to join Gell-Mann, who was eager to develop a gauge theory for the strong force, which was believed to be mediated by pions. From the fact that the massless vector bosons predicted by the *exact* gauge theory were absent, they postulated that the symmetry of the strong force was *not exact*. Since the symmetry of a gauge theory is a Lie group, Cartan's analysis of simple Lie groups served to classify types of gauge theory, so they selected those suitable for the strong force for further investigation. In January 1961, in a moment of epiphany, Gell-Mann devised a way of classifying particles in terms of their electric charge and strangeness. The two nucleons, the lambda, the three sigmas and the two xi particles, formed an *octet* in this abstract space constituting four isotopic spin multiplets. Similar octets were evident for the spinless (spin = 0) and vector (spin = 1) mesons. The same symmetry was simultaneously and independently discovered by Yuval Ne'eman, an Israeli military officer pursuing his doctorate under Salam in London. Although Ne'eman's paper was the first into print, Gell-Mann's reputation ensured that his paper drew more attention, and so the symmetry is known by his term: the '8-fold way'. The holy grail of particle physics, a 'higher' symmetry by which to classify the profusion of particles, seemed finally to be at hand.

Gell-Mann and Glashow discovered that the baryons were a supermultiplet, and would all have the same mass under an exact symmetry corresponding to the unitary group in three dimensions. However, breaking the symmetry split the supermultiplet into a number of isotopic spin multiplets. The mass splitting between multiplets is significant because the unitary symmetry, being *badly broken*, is far from exact. The splitting within each multiplet is narrower because the isotopic spin symmetry is an *almost exact* symmetry, and is a consequence of electromagnetism not conserving isotopic spin (protons and neutrons would be indistinguishable if electric charge did not exist). While the isotopic spin symmetry corresponded to the unitary group in two dimensions, the 'higher' symmetry was this same group in three dimensions. Interpreting this symmetry as gauge invariance, Gell-Mann and Glashow pondered the identity of its meditating vector bosons. In early 1960, J.J. Sakurai in Chicago had rejected the presumption that the pion was an elementary particle, and had posited that the strong force was due to a gauge invariance that conserved isotopic spin, strangeness and baryon number. The issue, of course, was that the theory predicted massless vector bosons as mediators, whereas the limited range of the force meant that this was not the case. Nevertheless, interest in the theory was stimulated by the discovery in 1961 of a triplet of rho mesons, each of which had about 75 per cent of the proton's mass and decayed into a pair of pions. The omega meson found several months later was even heavier, and decayed into a trio of pions. The phi meson was heavier still, and decayed into two kaons. Since the theory involved a vector field for each component symmetry of the invariance, it was speculated that these mediators might be the rho, omega and phi. Unfortunately, Gell-Mann and Glashow found that Sakurai's gauge theory did *not* correspond to a simple Lie group but a combination of several, and it could be generalised as a group corresponding to the '8-fold way'. Although it was a mystery how they could have acquired their masses, Gell-Mann and Glashow wondered whether these vector mesons might correspond to the massless vector bosons of the gauge theory. However, when it was found that the rho, omega and phi mesons did *not* play a significant role in nuclear processes it became evident that they were *not* mediators. Frustratingly, therefore, while there was no sign of the bosons predicted by the theory, there was a profusion of vector mesons that it could not explain.

When Enrico Fermi used the new synchrotron in Chicago in 1952 to smash pions into nuclei to investigate how the strong force worked, he was surprised to discover four heavy baryons with spin $= 3/2$ and electric charges ranging from -1 to $+2$. These were interpreted as a pion 'flirting' with a nucleon, circling it for 10^{-23} second before withdrawing, giving the momentary appearance of a single particle with a mass that was precisely the sum of the pion and the nucleon. The striking delta-shaped peak in the energy profile led to these being named 'delta' particles. A pion is spin $= 0$, so the spin was presumed to be the $1/2$ spin of the nucleon added to the spin represented by their orbital motion. The $+2$ electric charge was seen as a positively charged pion flirting with a proton, and the -1 as a negatively charged pion and a neutron. It was an ingenious theory.

At a meeting at CERN in Geneva in July 1962, a trio of resonances of the sigmas and a pair of resonances for the xis were announced. The sigmas had a strangeness charge of –1 and the xis had strangeness –2. The fact that they were spin = 3/2 implied that they were heavy relatives of the delta quartet. Gell-Mann and Ne'eman were in the audience, and when the speaker invited questions, Gell-Mann – sitting as usual in the front row – was the first to be acknowledged. Interpreting the new particles in terms of a decimet, he predicted a particle of isotopic spin 0, electric charge –1, and strangeness –3, which would complete this group, and he proposed that it be named omega. Furthermore, he pointed out that it should be able to be produced by firing very energetic negative kaons at protons and that, because it would have a relatively long lifetime, it should leave a detectable track in a bubble chamber. Ne'eman, who would have made the same prediction if he had been given the chance, held his peace. In mid-December 1963, Nicholas Samios headed a Brookhaven team in a search for this particle using a large new bubble chamber. They did not have to wait long: it turned up a month later.

How things had changed since 1960! By the end of 1962, all the known mesons could be assigned to singlet or octet symmetry groups, and the baryons assigned to

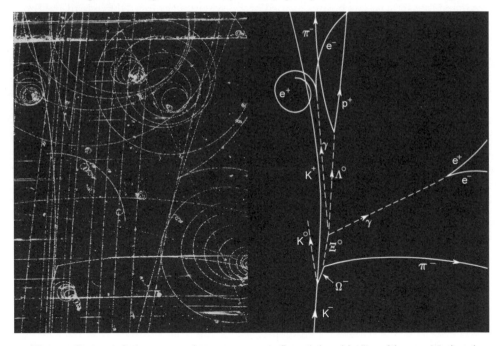

The prediction of the omega baryon was confirmed by this Brookhaven National Laboratory bubble chamber image (left). The interpretation (right) shows a negative kaon entering at the bottom and interacting with a proton to produce a neutral kaon (whose track could not be detected and so is drawn dotted), a positive kaon and the negative omega. After a short period, the omega decayed into a negative pion and a neutral xi, which in turn decayed into a neutral lambda and two gamma rays (which later reveal themselves by each producing an electron–antielectron pair whose tracks diverged in a characteristic manner). In due course, the lambda decayed into a proton and a negative pion.

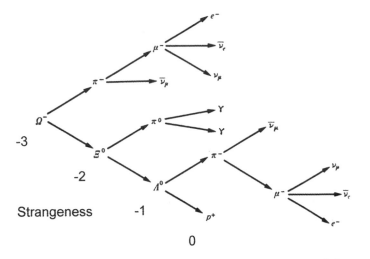

This decay tree depicts the way in which the triple-strangeness of the omega can be progressively shed via doubly-strange xi and singly-strange lambda baryons, before leaking away completely in the formation of the proton.

singlet, octet or decimet groups. Nevertheless, there was scepticism concerning how these symmetries should be interpreted in terms of the strong force.

QUARKS

By 1963, Gell-Mann, like Sakurai, had concluded that the hadrons could not be fundamental, they had to be composites of some as yet unobserved class of particle. Presuming that the '8-fold way' held a deep *physical* significance, he set out to infer from the various symmetry groups the properties of the constituent particles, which he whimsically named 'quarks'.

The first question was how many quarks a baryon contained. This was an issue of spin. With integer spin, bosons could combine only to form bosons, but fermions, having half-integer spin, could form either fermions or bosons depending on whether they combined in odd or even numbers, respectively. As fermions, baryons therefore had to have *three* quarks. As to the number of 'flavours' of quark, two were required to account for the range of electric charges (–1, 0, +1 and +2) with a third for strangeness. Gell-Mann named the 'normal' quarks 'u' (for 'up') and 'd' ('down') and the strange quark 's'. As a convention, he made a baryon with an electric charge of +2 from a trio of 'u', which implied that the charge on this quark was +2/3. A baryon with a charge of –1 was a trio of 'd' quarks with a charge of –1/3. The 's' had to have an electric charge of –1/3 because the omega baryon (which had recently been predicted but had yet to be confirmed) had a strangeness of +3 and an electric charge of –1. This explained why there were no baryons with an electric charge of –2. Also, depending upon how the spins of the quarks were aligned, baryons with spins of 1/2 and 3/2 were possible but a spin of 5/2 was not, which explained why such baryons were absent. Regarding the mesons, he reasoned that because they were bosons they had to comprise *pairs* of quarks. This explained why the mesons were lighter than the baryons. In fact, the 'u' and 'd' quarks were found to have similar masses, while the 's' was significantly heavier, which explained why strange baryons were heavier than nucleons, why the kaon was heavier than the non-strange mesons, and why the omega was the heaviest of all. The spin = 3/2 resonances that had been thought to be pions 'flirting' with spin = 1/2 baryons were now realised to contain quarks which were in excited vibrational states in much the same manner as the

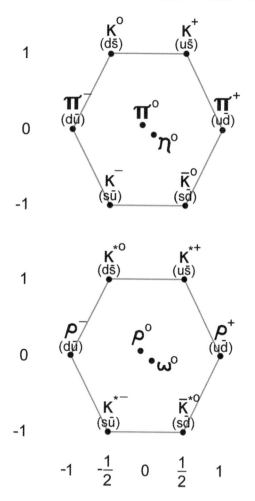

These octets depict the symmetry groups of the spin = 0 (upper) and spin = 1 (lower) mesons in terms of quarks and antiquark pairs. The horizontal scale is derived from isotopic spin and the vertical scale is strangeness. The eta and neutral pion, and the neutral rho and omega mesons are differing 'combinations' of the 'u' and 'd' quarks and their antiquarks.

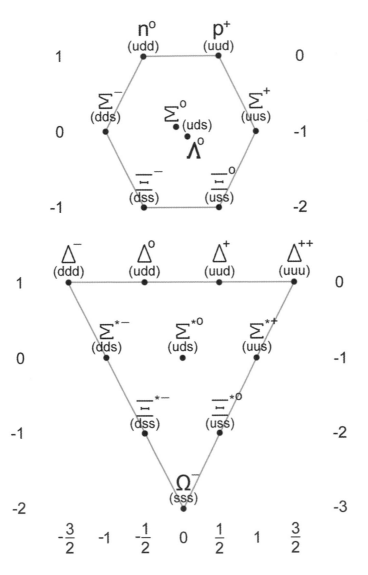

These symmetry groups show the relationships between the spin = 1/2 (upper) and spin = 3/2 (lower) baryons in terms of quarks. The horizontal scale is derived from isotopic spin, the right-hand scale shows strangeness and the left-hand scale shows hypercharge (strangeness plus baryon number).

atoms of a molecule. In retrospect, it was evident that the pion was a spinless (spin = 0) meson because the spins of its constituent quarks were *opposed*; if the spins were *parallel*, the result was a spin = 1 vector meson.

Once baryons were known to consist of three quarks, the cause of isotopic spin symmetry became apparent: the different flavours of quark had different masses and their combinations resulted in slightly different masses within the broad band of the multiplet. Although isotopic spin was conserved in strong interactions, it turned out *not* to be a fundamental quantity, which belatedly revealed that the efforts to devise a gauge theory based upon its invariance had been doomed to failure.

After years of frustration trying to make sense of the hadrons, the quark model was attractive. It explained why baryons formed singlets, octets and decimets, and why mesons formed only singlets and octets. The fact that it was derived from a consideration of symmetry was appealing. Nevertheless, in light of the absence of observational evidence for its *fractional* electric charges, it appeared to a community leaning towards the bootstrap theory to be a futile exercise in reductionistism.

FEYNMAN'S PARTONS

In 1952 Robert Hofstadter at Stanford decided to utilise high-energy electrons to investigate the internal structure of nucleons in a manner reminiscent of the way that Rutherford had used alpha particles to probe the structure of the atom (during which he had discovered the nucleus), except that the tiny electrons could probe much finer structure. The way in which the electrons were scattered revealed that nucleons were *not* the point-like objects that had been thought; in fact their electric charge appeared to be 'smeared out' over a spherical volume 10^{-15} metre across, prompting the proposal that they were *clouds of particles*, perhaps (in the thinking of the time) virtual mesons.

Accelerators of the early 1960s were not powerful enough to determine whether there were *quarks* inside hadrons, but the two-mile-long Stanford Linear Accelerator (SLAC) commissioned in 1967 could accelerate electrons to almost the speed of light and slam them into protons at energies approaching 20 GeV. In an attempt to knock out a quark, Jerome Friedman, Henry Kendall and Richard Taylor started firing electrons at protons. In August 1968 R.P. Feynman of Caltech paid a surprise visit

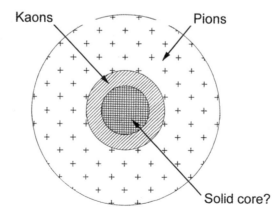

Before nucleons were shown to comprise trios of quarks, one theory was that they were clouds of mesons surrounding some form of dense core.

to SLAC to "snoop around". His friend James Bjorken was away that day but his postdoc, Emmanuel Paschos, showed Feynman the latest scattering data. Upon being shown a graph that had baffled the experimenters, Feynman fell to his knees, clasped his hands in front of his face and thanked the Lord for presenting such a spectacular verification of an idea that he had been working on for the last few months. The proton was in fact a cluster of point-like particles, which he dubbed 'partons'. Although the quark had not been detected *directly*, Bjorken's calculations implied that there were electrically-charged particles 'jiggling' within the volume occupied by the proton. The manner in which they scattered electrons indicated that they were not clumped together, but were free to move *independently* within the confines of the proton. Nevertheless, they were evidently so tightly bound together that they could not be knocked out. Physicists dissatisfied with the bootstrap theory of hadrons seized upon partons as proof that quarks were real. This identification was reinforced by SLAC experiments in 1971 showing that the constituents of the proton were fermions. The following year, CERN reported that neutrino scattering off protons observed by the 20-tonne French-built Gargamelle bubble chamber had revealed evidence of *fractional* electric charges. As these discoveries were being made, the nature of the weak force was being resolved.

THE MASS ISSUE

Glashow's attempt to formulate a gauge theory to unify the electromagnetic and weak forces had been stymied by the fact that it had predicted *massless* vector bosons as force carriers – and one could be identified as the electromagnetic photon – but the limited range of the weak force meant that its mediators could *not* be massless. In 1960, however, while studying spontaneous symmetry breaking in quantum fields, Jeffrey Goldstone, a physicist at Cambridge, discovered that the process produced massless *scalar* bosons. The traditional Mexican sombrero hat possesses a rotational symmetry, but if a ball is thrown in and comes to rest in the depressed annular brim, the presence of the ball spontaneously breaks the underlying rotational symmetry of the hat.

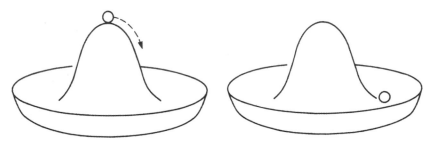

The rotational symmetry of a Mexican sombrero is lost when a ball perched on its peak slips off and settles in the annular brim. The Goldstone boson plays this same role in spontaneously breaking the symmetry of a scalar field.

This shape also describes the function of the potential energy of a quantum field with an *internal* symmetry. An internal symmetry is based on the *intrinsic* properties of the particles rather than on their position and motion in space. The particles of such a symmetry are interchangeable in some abstract space by the rotation of a 'selector arrow'. The spontaneous breaking of the symmetry of a quantum field is known as the Goldstone theorem. The object that breaks the exact (or 'perfect') symmetry is a *Goldstone boson*. Wherever the boson settles on the minimum-energy circle (which is the counterpart of the hat's annular brim) it is free to 'oscillate' with an *arbitrarily small* amount of energy, and therefore must manifest itself as a *massless* particle, for otherwise its rest mass would establish a limit below which its energy could not fall. In 1962 Goldstone wrote a paper which attracted little attention because massless scalar bosons seemed to be a mathematical artefact of no physical significance. In 1964, however, P.W. Higgs at Edinburgh University discovered that the massless scalar bosons from the spontaneous symmetry breaking would 'combine' with and *bestow mass upon* the massless vector bosons of the gauge theory. The process was analogous to the phase change of spontaneous magnetisation in which the profile of energy versus magnetic field transitioned from a U-shaped to a W-shaped profile as the minimum energy state adopted a non-zero magnetic potential. The process was different as the Higgs field froze, but the effect was similar. In its 'hot' state, the field strength was zero (that is, the ball was atop the sombrero hat) but this symmetry was broken as the field 'froze' and adopted a non-zero value. In magnetisation the released energy of angular momentum was transferred into the magnetic field, locking it away so that it could not be shed by further cooling. A particle is an *oscillation* in an energy field, and is a boson or a fermion depending upon whether the field is a vector or a spinor field. When the *scalar* Higgs field froze, it 'biased' the related fields. The minimum energy of a particle is its rest mass. By giving particles energies that they could not shed, the freezing Higgs field gave them masses.

The realisation that the spontaneous freezing of a scalar field would bestow mass on particles was an epiphany, but Higgs thought it an intriguing mathematical device

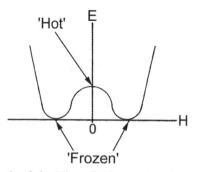

In its 'hot' state the strength of the Higgs field was zero, but as the field 'froze' and adopted a non-zero value this symmetry was spontaneously broken. Although there are two frozen values, there is only one solution because in the case of a scalar field only the magnitude is relevant, rather than sign in a vector field.

lacking physical significance. Even Glashow, who knew Higgs and really should have made the connection, thought it a "terrifically looney" idea. The first draft of Higgs's paper, 'Broken Symmetries and the Masses of the Gauge Bosons', was rejected because quantum field theory was in the doldrums and the paper was deemed uninteresting.* The essence of the work – the significance of which was not fully appreciated at the time – was that a symmetrical theory can produce an *asymmetrical* solution.

WEINBERG'S SYNTHESIS

After gaining his doctorate from Princeton, Steven Weinberg became an itinerent postdoc with periods at Columbia, Berkeley and Harvard leading to MIT in 1967. On his travels, he had met Abdus Salam who, having belatedly accepted Glashow's view on gauge theories, had attempted in 1964 with J.C. Ward to unify the electromagnetic and weak forces but, like his predecessors, had been defeated by the issue of the masses of the mediating vector bosons. In 1960 Glashow had speculated upon the existence of a 'neutral current', and in his 1961 collaboration with Gell-Mann had argued for a 'strangeness-changing neutral current' – a process for which there was no evidence.

Gell-Mann's quark model explained strange hadrons in terms of a specific quark. When the weak force changed the flavour of one of the quarks in a strange hadron, the process was *strangeness-conserving* if it did not involve this strange quark and *strangeness-changing* if it did. A negative kaon (which is a strange meson) had been observed to decay into a neutral pion, an electron and its associated antineutrino. This was a *charged current* because the electric charge transferred from the hadron to the electron, which is a lepton. Gell-Mann and Glashow noted that a gauge theory also required a *neutral current* in which the electric charge was taken by the hadronic rather than the leptonic decay products. This would occur if a negative kaon decayed into a negative pion. They took the absence of evidence for the strangeness-changing neutral currents as indicating a flaw in the theory, and abandoned it.

In 1967 Weinberg realised that the Higgs mechanism explained the masses of the vector bosons mediating the weak force, and by incorporating the Higgs mechanism into the Glashow–Salam–Ward gauge theory he unified the weak and electromagnetic forces in an 'electroweak' theory. Weinberg's initial motivation had been simply to supercede Fermi's theory with one using exchange particles, the fact that the result encompassed electromagnetism came as a considerable surprise. At the high energies prior to the symmetry of the scalar field being spontaneously broken, all four vector bosons of the gauge theory were massless. The photon of the electromagnetic force remained massless after the symmetry broke, but the scalar

---

* The Higgs mechanism was independently discovered by P.W. Anderson of Bell Laboratories while on sabbatical at the Cavendish Laboratory in Cambridge in England in 1962, and by Robert Brout and François Englert of the University of Brussels in 1964, without the various players being aware of one another's work.

bosons that broke the symmetry bestowed mass on the others. The electromagnetic and weak interactions appear so different because, in the broken symmetry, the theory has an *asymmetric solution*. It came as a great surprise to realise that the symmetry was *still present in the theory*. Weinberg has described the realisation that a symmetric theory can produce an asymmetric solution as "one of the greatest liberating developments of twentieth-century science". It is hard to imagine, but the ubiquitous electron and the elusive neutrino had similar properties under the exact symmetry.

Abdus Salam, since 1964 the director of the International Centre for Theoretical Physics at Trieste in Italy, independently reached the same conclusions in 1968, but for several years he and Weinberg remained unaware of each other's work. After an initial period of excitement, their enthusiasm waned due to their inability to establish that the theory upheld local gauge invariance and was therefore renormalisable. The theory was met with indifference. In fact, Weinberg's paper was not even cited until

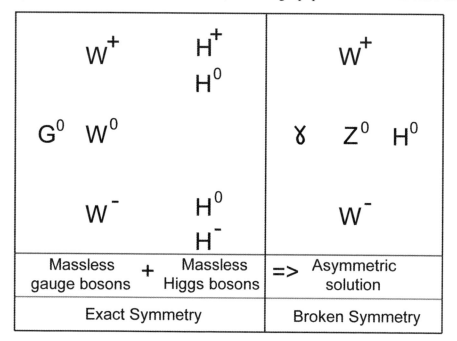

The electroweak theory involved four Higgs fields which, being spinless (spin = 0), were described by the Klein–Gordon equation. Under the exact symmetry, the gauge theory's four massless vector bosons (the W trio and the proto-electromagnetic field G°) were indistinguishable. The Higgs mechanism mixed the initial massless charged weak fields with the similarly charged scalar Higgs fields to give the W fields mass, and the two massless neutral vector fields mixed with the neutral Higgs fields to give the massive neutral weak Z field, the massless neutral electromagnetic field, and the massive neutral Higgs field. The quantised fields manifested a trio of massive vector bosons (the $W^\pm$ and Z°), a massless vector boson (the photon) and a massive neutral scalar boson (the yet to be identified Higgs boson, H°).

1970, and then only once, and by then both men had set the theory aside and moved on to other work. Although the Gell-Mann–Feynman theory (as extended by Nicolo Cabibbo in Italy) explained all known weak interactions, and the electroweak theory explained no mysteries, the new theory was more appealing because it *required* the observed dependence of the force on the spins of the particles. Weinberg was thus able to predict the relative strengths of the charged and neutral currents, and enable the experimenters to calculate the sensitivity that their apparatus would have to achieve in order to stand a chance of detecting the neutral current.

A profound implication of Weinberg's discovery that the electroweak symmetry was spontaneously broken in a *phase change* was that the Universe had once been *incredibly hot*, and had cooled. In 1967 the idea of the Big Bang was only just gaining credibility, and its opponents, who thought that the Universe had no beginning, rejected primordial phase changes. Nevertheless, the realisation that J.C. Maxwell's unification of the varied phenomena of electricity and magnetism had not been a one-off, and that electromagnetism and the weak force that controlled subatomic decay embodied a hidden symmetry, raised the prospect that *all* the forces of nature were once unified, and that a succession of phase changes introduced the distinctions that we perceive. It was a heady thought!

BREAKTHROUGH

The infinities that plagued Weinberg's unified gauge theory were a manifestation of the manner in which the force was calculated by summing the probabilities of all the ways in which a specific interaction could occur, because only charged currents were taken into account. In late 1969, Glashow invited John Iliopoulos and Luciano Maiani to Harvard as postdocs, and together they tackled the strangeness-changing neutral currents issue. In 1964 Glashow and James Bjorken had speculated on aesthetic grounds (in order to match the numbers of quarks and leptons) that there was a *fourth* quark, but the idea attracted no interest. Reviving it, they noted that while this new quark introduced another way for the weak force to generate a neutral current, the probabilities of the different interactions cancelled out to 'suppress' it.

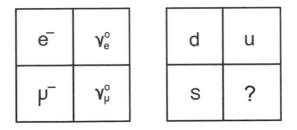

On the presumption that there were matching numbers of quarks and leptons, in 1964 S.L. Glashow and James Bjorken interpreted the gap in this pattern to mean that a fourth quark remained to be found.

More significantly, a fourth quark made it feasible to *fully describe* the weak force, including its hadronic decays.

Although Glashow had collaborated with Weinberg on several papers in the late 1960s, he was unaware of the electroweak gauge theory, and when he gave a seminar at MIT in early 1970 on how the fourth quark suppressed the strangeness-changing neutral current, Weinberg did not pick up on the fact that it would also complete his partial theory, which had addressed only leptonic decay. After Maiani returned to Italy, Iliopoulos attempted to prove that the unified theory was renormalisable, but without success, and he then resumed the postdoc trail. While at the University of Marseilles in France in 1972, he found that quarks *and* leptons must each occur in *complete families* in order to yield the internal consistencies required for their gauge theories to be renormalisable – an insight that would deliver a dividend a few years later.

Meanwhile, Gerard 't Hooft, working for his doctorate under the supervision of Martin Veltman at the University of Utrecht, had reinvented the Goldstone theory, the Higgs mechanism, the Weinberg–Salam theory, and the need for a neutral current and, by establishing that the infinities introduced by charged-current interactions were cancelled by those from the neutral current, had proved the electroweak theory to be renormalisable. In fact, he demonstrated that *all* gauge theories for quantum fields with spontaneously broken symmetries were intrinsically renormalisable. By mid-1971, therefore, all the pieces of electroweak unification were in place. Nevertheless, it would remain just an elegant hypothesis until evidence was found for the new particles.

Apart from identifying the $W^\pm$, which mediated the weak charged current, the key would be to find the rarer neutral vector boson, Z°. One search technique was to fire neutrinos at hadrons. In a typical charged-current interaction, a neutrino would strike a quark inside a hadron and exchange a $W^\pm$ with it, changing the flavour of the quark and turning the neutrino into a muon, which took the hadron's electric charge. In the hypothetical neutral current, the neutrino would retain its identity and the electric charge would be carried away by one of the hadronic by-products. While distinctive, such interactions would be elusive, which was probably the reason they had not been observed in the course of previous experiments. Until they could be confirmed, there was the prospect that they were not seen because the theory was flawed, and they did not occur. The first sign of this neutral current was spotted in 1973 by F.J. Hasert observing interactions between neutrinos and neutrons in CERN's Gargamelle bubble chamber. Although the Z° had not been observed directly, there was celebration because the observation of its effect was the first experimental support for the electroweak theory. Hot on the heels of proof of the renormalisability of gauge theories, this revived interest in quantum field theories. The Glashow–Iliopoulos–Maiani results were generalised in 1975 by Makato Kobayashi and Toshihide Maskawa in Japan to fully describe the interplay between the strong and weak forces, with the weak altering the flavours of *both* leptons and quarks within hadrons. To finally prove the electroweak theory required specifically observing the $W^\pm$ and Z° mediators, but the *production* of these heavy particles was beyond the energies available in the early 1970s. However, the introduction of the proton–antiproton colliders at CERN and Fermilab a decade later opened the door to 100 GeV energies. CERN's machine entered service in early 1981 and spotted its first $W^\pm$ in late 1982. The team leader, Carlo Rubbia of Harvard, told a workshop in Rome on 13 January 1983 that after sifting through millions of bubble chamber pictures in search of tracks that could *only* have been left by weak bosons, they had spotted *four* cases of W^- decay and one of W^+ decay, putting the $W^\pm$ mass at 81 (± 5) GeV, in excellent agreement with the prediction of 82 (± 2.4) GeV. The Z° was identified several months later. Its greater mass of 95 GeV was also consistent with the theory. Reflecting the enormous scale of these experiments, each report had in excess of 130 authors.

The outstanding 'loose end' was to verify that it was the Higgs mechanism that spontaneously broke the electroweak symmetry. Many properties of the spinless Higgs boson were specified by the theory, but not its mass. Early estimates based on circumstantial factors spanned the range 10–300 GeV, although it might be as high as 1 TeV. Contemporary accelerators could search only the lower part of this range. The primary objective of the Superconducting Super Collider that was to have been built in Texas was to confirm the existence of the Higgs boson, but this accelerator was cancelled in the early 1990s. This task will now fall to the Large Hadron Collider that CERN hopes to introduce by 2010. If the Higgs boson proves illusory, then at best our understanding of the Higgs mechanism is flawed, and at worst it has no physical significance. In a sense, therefore, there is an alarming parallel with the belief a century ago that electromagnetic waves could not exist in a vacuum, they had to propagate through *something*, and hence it was inferred that an 'aether' pervaded space – an inference that proved false. *Something* broke the electroweak symmetry. If it was not by the Higgs mechanism, then it was by some as-yet-unidentified agent. Either way, it is important that this issue be resolved.

THE DISCOVERY OF THE CHARMED QUARK

In 1962, six years after C.L. Cowan and F. Reines finally found the neutrino, Melvin Schwartz, L.M. Lederman and Jack Steinberger at Columbia discovered that the neutrinos from pion decay were *different* from those emitted by unstable nuclei. In fact, Julian Schwinger at Harvard had predicted that there *would* be two types, with the weak force turning an electron into an electron-neutrino and vice versa, and a muon into a muon-neutrino and vice verse, but with no mixing. This imposed a symmetry among the leptons, with two pairs. The discovery of the neutron and the prediction of the neutrino in the early 1930s had prompted the belief that there might be a 'balance' between the numbers of leptons and baryons, with two of each. When the muon was found in 1937 it was presumed to be a meson, but a decade later was realised to be a lepton. The balance was temporarily restored by the strange lambda baryon in 1951, but the idea of such a correspondence was undermined by the profusion of hadrons revealed by the new accelerators in the 1950s. Following Gell-Mann's proposal that hadrons were made of quarks, it was the prospect of a balance between the numbers of quarks and leptons that prompted Glashow and Bjorken to speculate in 1964 that there might be a fourth quark, which they whimsically named the 'charmed' quark. If this existed, then there ought to be a whole new family of heavy hadrons waiting to be discovered, but testing this hypothesis was not possible using the accelerators of the time. Although accelerator energies increased in the 1960s, no evidence appeared to suggest such particles. On the other hand, no one had conducted a search. In any case, when something unusual was observed, the possibility of it being a new quark was never the experimenter's initial hypothesis. Consequently, when something did appear it was not immediately recognised. Carlo Rubbia of Harvard saw some unusual results in a neutrino-scattering experiment at Fermilab in late 1973, but he wrote them off as an instrumentation artefact. Eighteen months later, however, a new quark was detected simultaneously by teams working at Stanford and Brookhaven.

In 1964, with the construction of SLAC underway, Burton Richter proposed an adjunct in the form of the Stanford Positron–Electron Asymmetric Rings (SPEAR). Instead of smashing the electrons from the linear accelerator into a target, they would be fed into a 'storage ring' in SPEAR and then a target would be inserted one-third of the way down the accelerator to enable the resulting antielectrons to be stored in a second ring. The two oval rings were magnetically manipulated to intersect in short straight sections so that the contra-rotating particles would hit head-on and particles would be created from the annihilation energy. Although SPEAR's 8 GeV 'energy budget' was not as great as the linear accelerator, by 1974 it was sufficient to produce 'charmed hadrons'. Meanwhile, Samuel Ting, a student of Martin Perl, was smashing 7 GeV protons into a beryllium target in Brookhaven's Alternating Gradient Synchrotron in order to create a heavy hadron that would decay into an electron–antielectron pair whose combined energies would yield the mass of the otherwise unobserved hadron. On 4 November 1974 Glashow received a call from Min Chen (a member of Ting's group) at MIT, requesting that he pay a visit to comment upon a prominent 'spike' in the energy distribution showing that

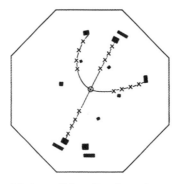

The discoverer of a new particle is traditionally allowed to name it, but the simultaneous discovery of the charmed quark posed a dilemma: Samuel Ting wanted to name the new hadron 'J', but Burton Richter, inspired by the pattern that the decay products left in his detector, preferred 'psi'. Lederman negotiated the name 'J/psi' (pronounced 'gypsy') as a compromise.

the experiment was creating a large number of particles at 3.1 GeV. After venturing that this was charmed, Glashow rushed back to Harvard to spread the good news, only to find his colleagues already in high spirits as a result of a report that SPEAR had spotted precisely the same thing! Richter was on sabbatical at Harvard, so he flew home to congratulate his team. It turned out that while Ting had observed the new particle decay into an electron–antielectron pair, the SPEAR team had made it by annihilating an electron–antielectron pair.

Although Glashow was confident that the charmed quark was involved, over the next few weeks other theorists developed their own interpretations of the data and so nine papers on this theme appeared in the first issue of *Physical Review Letters* of the new year. As modelling of a series of similar particles in 'excited' states would show, this J/psi was a meson with charm–anticharm quarks – a charm-neutral pairing dubbed 'charmonium'. Later in 1975, Ting's group fired neutrinos at protons to make charmed quarks 'one at a time' as the weak force altered the flavour of one of the constituent quarks. After examining a million bubble chamber pictures they found a single event which perfectly fitted the prediction for a baryon with 'naked charm'. Mesons with naked charm followed: the D-mesons were a 'c' quark paired with an 'u' or a 'd', and the strange F-meson was a 'c' with an 's'. The existence of the charmed quark not only reinforced Gell-Mann's quark model, it also balanced the numbers of quarks and leptons, with four of each, which hinted at a deep symmetry between these two classes of particle.

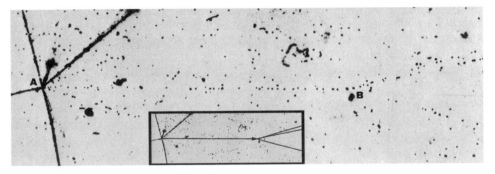

The first case of 'naked charm' was identified in 1976 in this photographic emulsion irradiated by neutrinos. The particle was produced by the interaction at point A and travelled left to right to point B, where it decayed after less than 10^{-12} second into three tracks.

COLOURED QUARKS, GLUONS AND ASYMPTOTIC FREEDOM

While the progress in unifying the electromagnetic and weak forces reinvigorated efforts to express the strong force as a gauge theory, there were significant problems with the quark model. In 1964 O.W. Greenberg of the University of Maryland had pointed out that while the Exclusion Principle prohibited more than one fermion occupying the same quantum state, a spin $= 3/2$ hadron comprising a trio of the *same* quarks would require at least two of them to have the *same* spin. One solution was to argue that fermions behaved like bosons when inside a hadron, but this was rather *ad hoc*. If the Exclusion Principle was to be upheld, a trio of 'identical' quarks must differ in some as yet unrecognised way. This property was required to have three possible values. However, the fact that this had no macroscopic manifestation implied that it, together with the quarks, was *confined* within a hadron. M-Y. Han at Syracuse and Yoichiro Nambu at the University of Chicago speculated in 1965 that the quantum number that distinguishes the otherwise identical quarks might be responsible for the force that confined them, although how *this* force related to the strong force was unclear. In late 1971 Gell-Mann teamed up with Harald Fritzsch, a newcomer to Caltech, to resolve this mystery. Drawing upon the analogy of the three primary colours combining to form a 'neutral' white, they argued that it was useful to think of quarks as coming in three colours and combining with each other such that hadrons were colour-neutral. It was evident that irrespective of their individual flavours, there had to be one quark of each colour in a baryon. In the case of the triply-strange omega baryon, therefore, one of the 's' quarks was 'red', one was 'green' and the other was 'blue'. They were not identical after all, and so were not in conflict with the Exclusion Principle. Also, since mesons had to be colour-neutral, the colours of the two constituent quarks had to cancel. This meant that a colour was paired with its anticolour (for example, 'red' with 'antired'). Although a meson comprised a quark and an antiquark, these did not have to be of the same flavour as only the colours were obliged to cancel. And since the only way to make colour-neutral hadrons from three colours was either to blend all three colours or to pair a colour with its anticolour, this explained why there were no mesons of four quarks, or baryons of five quarks.

Although this 'colour scheme' was initially presented merely as a mathematical model, it was soon realised that (as Han and Nambu had speculated) it generated the force that confined quarks within hadrons. The manner in which SLAC's electrons were scattered by hadrons showed that, in addition to the electrically charged quarks, the hadrons contained electrically neutral objects which were evidently the 'gluons' that mediated this force. In 1973, in a paper entitled 'Advantages of the Colour Octet Gluon Picture', Gell-Mann, Fritzsch and Hans Leutwyler presented an SU(3) octet of massless spin = 1 bosons as the coloured counterparts of the electromagnetic photon.

Although gluon exchange can be thought of simply as the transfer of a colour, the situation is actually more complex. If a green quark exchanges a gluon that turns a nearby quark green, what colour should the original quark become? In fact, a gluon is better thought of as a combination of fields. With three colours, there are *nine* combinations. However, matched solutions such as red–antired are 'white' and therefore play no role, and in accordance with the rules of group theory,* there are two 'off white' combinations in addition to the six coloured solutions blending a colour and an anticolour, and so the force is mediated by *eight* gluons.

---

*   An SU(n) gauge theory has n^2-1 vector bosons as mediators.

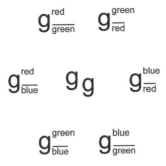

In addition to the easy to represent combinations combining pairs of hues, the octet of gluons contains two 'off white' combinations of the three colour fields.

A quark is a combination of all three colour fields. A quark is green because it has an absence of antigreen. Hence, when a quark emits a green gluon, the colour of the quark changes to compensate for the *other* shade of the gluon. As a result, emitting a green–antired gluon will leave the quark red, and emitting a green–antiblue gluon will leave it blue. The gluon emitted when a quark changes colour is therefore a mixture of the initial colour and the opposite of its new colour. Another way to consider this is to say that if a red quark turns blue by emitting a gluon, the gluon takes away the red and antiblue and leaves behind the blue. However, while such analogies provide some insight into how this force operates, to fully understand it requires a mathematical formalism.

In 1970 Gerard 't Hooft in Holland laid the basis for the discovery of *asymptotic freedom*, the property of the colour force which confines quarks and gluons inside hadrons. Its details were refined simultaneously in early 1973 by H.D. Politzer, a student of Sidney Coleman at Harvard, and by D.J. Gross and his student Frank Wilczek at Princeton.

The Uncertainty Principle permits pair-production in the vacuum as long as the particles annihilate within the ΔT interval. Each 'real' electrically charged particle is immersed in a cloud of virtual charged particles which are either attracted or repelled by the steep gradient in the electric field. Although the *individual* virtual particles flit in and out of existence, the production and annihilation is ceaseless and the cloud of electricity *polarises* the vacuum. The 'bare' charge on an electron is extremely large (possibly even infinite) but this 'shielding' masks most of the bare charge and the electric charge that we *measure* is merely the leakage. As their shielding is identical, all electrons have the same 'effective charge' at a given range. The repulsion between two similarly charged particles varies inversely with their separation. In a *long distance* encounter in which they do not significantly penetrate each other's shields, they behave as if each had a *unit* of electrical charge. But the situation is more complex in a high-energy collision in which the particles penetrate one another's shields, as *two* issues must be considered – one involving the bare charges on the real particles and the other involving the shielding of their clouds of virtual particles. At the very short ranges of deep penetrations the true magnitude of the bare charge

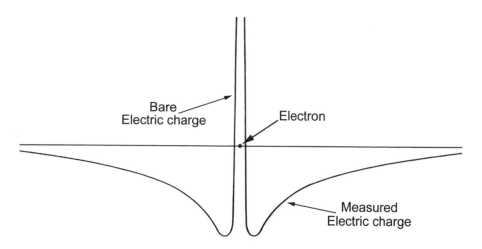

The 'bare' charge on an electron is extremely large (possibly even infinite) but most of it is 'shielded' by vacuum polarisation and the electric charge that we *measure* is merely the leakage.

begins to reveal itself, and the strength of the electromagnetic force *increases*. A clue to quark confinement came with the discovery that the strength of the colour force *decreases* with deeper penetrations.

In fact, the vacuum is similarly polarised by *other* fundamental charges. The bare colour charge of a real quark is shielded by a cloud of virtual quark–antiquark pairs; for example, green–antigreen pairs will form an antigreen shield around a green quark, masking the intensity of its hue. Although this would tend to imply that the colour force should possess the same characteristics as the electromagnetic force, this is not the case because the gluons of the colour force are themselves coloured, and (in contrast to photons, which are electrically neutral) gluons feel the force that they mediate. Furthermore, the different types of gluon can interact with each other. In addition, a gluon can split spontaneously, with its colour mix being shared between the two resulting gluons, with one taking the *third* colour and the other its anticolour – for example, a red–antigreen can split into a red–antiblue and a blue–antigreen. Consequently, if a red quark emits a red–antigreen gluon that gives rise to a sequence of splittings, a *multitude* of virtual gluons bearing the quark's colour will establish a cloud that distributes that colour over a volume of space much larger than the size of the quark (which is point-like) *magnifying* its colour. This does not occur in the case of an electrically charged particle because the virtual photons do not smear out its electric charge.

The effective strength of the force is very strong *outside* the radius of the volume occupied by the cloud of virtual gluons because the magnification of the colour of the central quark more than compensates for the shielding of the virtual quarks and antiquarks in the polarised vacuum. A deeply penetrating collision will sense *fewer* gluons, however, and so at high energies the effective strength of the force decreases. As the quarks inside a hadron tend to reside well within one another's gluon clouds,

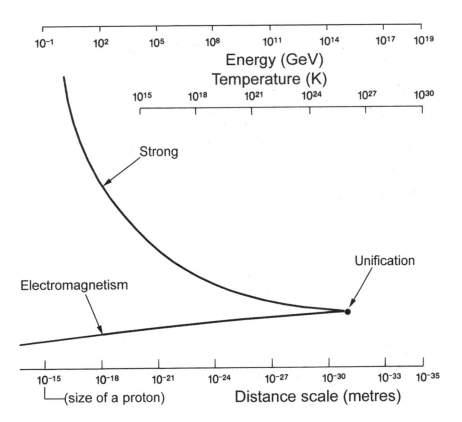

At the very short ranges of deep penetrations the true magnitude of the bare charge begins to reveal itself, and the strength of the electromagnetic force *increases*. In the case of the colour force, however, the strength *decreases* with deeper penetrations. This raised the prospect of a unification at some exceedingly high energy.

the force acting on them is small, and tends to zero the closer they drift. This is why a trio of quarks can sit essentially immobile alongside each other, and also why, for example, in the 'charmonium' meson the quark and antiquark do not annihilate each other – the force of attraction is so slight that they are sluggish. Quarks behave as if they are *un*bound and yet are confined within hadrons. This property was dubbed asymptotic freedom because they become *free* as the separation between quarks tends to zero.

The effective strength of the colour force increases until about 10^{-15} metre (the diameter of a nucleon), after which it declines. The force is so strong because it has eight mediators. Although hadrons are 'white', the colour fields from the individual quarks create a *residual* attraction between hadrons and it is *this* that manifests itself as the strong force. It was therefore belatedly realised that the force that binds nucleons is analogous to molecular bonding: although atoms are electrically neutral, their electron clouds are distorted when they come into proximity and the resulting electrical force forms molecules.

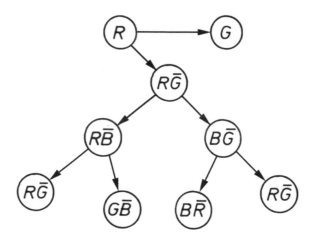

Because the gluons that mediate the colour force are themselves coloured, they can interact with each other. As a result, if a red quark emits a red–antigreen gluon that gives rise to a sequence of splittings, the quark becomes surrounded by a *multitude* of virtual gluons that bear *its* colour, thereby *magnifying* it.

Electron scattering in *very deep* penetrations into nucleons was described fairly accurately by the perturbation techniques developed for quantum electrodynamics, where the only complication is the shielding effect. (Indeed, Feynman – who had played a role in developing these techniques – utilised them to interpret the SLAC data as indicating that the objects inside a proton behaved as if they were free.) However, such techniques failed at distances at which the force was so strong that it confined the quarks, and to discuss confinement physicists initially had to rely on 'handwaving' arguments and analogies – such as thinking of quarks as being joined by elastic. Quarks cannot be liberated individually *because* when one is struck by an electron, the imparted energy drives it out of its partners' virtual gluon clouds, the *increasing* force provides the energy for the production of a *real* quark–antiquark pair, the new quark is drawn inside and the antiquark joins with the emerging quark; the particle that then emerges is therefore a *hadron* rather than an isolated quark. If the effective strength of the force *decreased* in proportion to the square of the range, as in the case of electromagnetism, then pair production would *not* be possible as the range increased, and the energised quark would emerge. Instead, as a quark strays from its partners, the colour force draws it back towards the locus at the centre of the hadron's spherical volume. In fact, hadrons have the sizes that they do precisely because their radius is the *furthest* distance a quark can stray before the rising energy in the colour field stimulates the hadronisation process. The collisional energy at which gluons 'lose their stickiness' corresponds to a temperature of 10^{14} K. Such energies are far beyond our capabilities, but if we could achieve them the colour force would become so feeble that hadrons would literally become 'unglued' and their quarks and gluons would spill out. This led to the profound implication that if the temperature of the early Universe was hotter than 10^{14} K, hadrons would have been unstable – in effect, they are a feature of a 'frozen' Universe.

QUANTUM CHROMODYNAMICS

In 1978, once the details of the gauge theory of the colour force had been refined, Gell-Mann named it 'quantum chromodynamics' in order to emphasise that it was a counterpart to quantum electrodynamics. It was belatedly realised that the efforts to devise a gauge theory for a force mediated by pions had failed because hadrons were composite particles. Unlike the spontaneously broken symmetry of the electroweak force, the symmetry of the colour force is *exact*. Consequently, all three colour fields are equally intense, and the *strength* of the force is indifferent to the specific colours of quarks. The fact that coloured quarks bound by coloured gluons obeyed a *hidden* symmetry explained a great deal. Although both the colour and weak forces influence quarks, they affect *different properties*. The *flavour* of a quark is not changed by the colour force, it is changed by the weak force. These forces act independently – in fact, if this were *not* the case it would be impossible to construct gauge theories that did not conflict. Nevertheless, there must be a deep relationship between the flavours of quarks and leptons, because both are subject to change by the weak force. Whatever this symmetry is, there is clearly a distinction because, as far as the colour force is concerned, the leptons are 'white'.

With the realisation that the Universe is a hierarchy based on quarks and leptons the reductionistic view was firmly established. The bootstrap philosophy that was popular in the 1960s had been a distraction. As the development of electrodynamics had moved Freeman Dyson to speculate in 1949, quantum field theory did indeed appear to be a description of *reality*. After decades of frustration, by the late 1970s physicists felt that they were finally making progress.

THE STANDARD MODEL

The structure of the atom was established in the 1920s, but another half-century passed before a comprehensive explanation of the nuclear realm was developed. A few years of synthesis had drawn many of the strands together and filled in some of the gaps. It had taken so long to separate the wheat from the chaff because the electroweak symmetry was well hidden and the symmetry of the colour force was confined within hadrons. Members of a symmetry group are interchangeable under transformations that can be likened to the 'arrow settings' on the selector in an abstract space. Quantum field theories are *unitary* symmetries. The U(1) of the electromagnetic force can be represented by rotation in a circle, the SU(2) of the weak force by a rotation in two dimensions and the SU(3) of the colour force by a rotation in three dimensions.

Although the electroweak theory unifies the electromagnetic and weak forces, the underlying symmetry is so badly broken that the force operates on a concatenation of the particle groups as $SU(2) \times U(1)$, each of which has its own bosons and coupling constant. This process of concatenation can readily be extended to include the colour force as $SU(3) \times SU(2) \times U(1)$. S.L. Glashow pointed out that because this explained almost all known particle interactions – with the exception of gravity – it effectively constituted a *Standard Model of Elementary Particle Physics*.

6

Seeking a theory of everything

UNIFICATION OF FERMIONS

When S.L. Glashow said that – with the exception of gravity – the concatenation $SU(3) \times SU(2) \times U(1)$ explained all known particle interactions and hence constituted a Standard Model of Elementary Particle Physics, he also observed that because there was evidently a balance between the numbers of leptons and quarks there might be a *fundamental* relationship between the electroweak and colour forces that would unite the two types of fermion. A unified theory of quarks and leptons should account for outstanding coincidences and mysteries. For example: Why are the electric charges on the electron and proton of identical magnitude? Why are the charges of the quarks quantised in units of one-third of this value? Why do the leptons and quarks possess the masses that they do? Why does the Standard Model need parameters such as the three coupling constants and the relative strengths of the components of the weak force to be derived empirically and then inserted into the equations? In the same manner that Einstein's theory of gravity was 'better' that Newton's because it *required* the inverse square law, a unified theory should explain *why* these parameters have such values.

In 1973, Abdus Salam and J.C. Pati of the University of Maryland ventured that 'lepto-quarks' were once interchangeable under a 'four-colour symmetry', and that a distinction was drawn between quarks and leptons when this SU(4) symmetry was spontaneously broken. Because the colour force emerged as an exact symmetry, they were effectively arguing that leptons are quarks with a neutral colour that the SU(3) gluons ignore. While this was an interesting idea, it posed a few problems, but before these could be investigated the theory was overtaken.

After receiving his doctorate from Yale, Howard Georgi returned to Harvard in 1972 as a postdoc to work with Glashow, and in early 1974 they proposed a *grand unified theory* utilising the simplest of E.J. Cartan's SU(5) symmetries to *contain* the individual members of the U(1), SU(2) and SU(3) groups, making the lepto-quarks interchangeable under transformations involving rotations in five dimensions. With Helen Quinn and Steven Weinberg now also at Harvard, Georgi developed the gauge theory. This predicted that there was a single coupling constant while the symmetry applied, and that as the energy density fell through 10^{15} GeV this was spontaneously broken and the coupling constants of the colour and electroweak forces diverged because the colour force was asymptotically free whereas the electroweak was not.

Of the 24 originally massless vector bosons of this grand unified force, eight could be identified with the massless gluons of the colour force and another four with those of the electroweak force (of which three later gained mass when the electroweak symmetry broke). The other 12, dubbed the 'X-bosons', were responsible for turning quarks into leptons and vice versa. In contrast to gluons (which are coloured and electrically neutral) and photons (which are both electrically and colour neutral) these bosons were both coloured *and* electrically charged. The 12 comprised a red, a green and a blue with charge $+1/3$, a similar trio with charge $-1/3$, and six others with charges of $\pm 4/3$. When the symmetry broke, the X-bosons acquired masses equivalent to the prevailing energy density – in fact, they became so 'superheavy' that they weigh as much as a human blood cell.

PROTON DECAY

The existence of X-bosons cannot be confirmed by producing them in collisions, because 10^{15} GeV is far beyond the capability of our accelerators. Nevertheless, just as observing the neutral current provided evidence for the heavy bosons of the weak force before they could be created in accelerators, there was the prospect of inferring their existence from an interaction that *required* their involvement. The weak force acts on both quarks *and* leptons, and by changing their flavours it manifests itself as 'decay'. The conservation of energy and of electric charge are insufficient to prevent a proton decaying into a neutral pion and an antielectron. To explain this apparent stability of the proton, physicists invented the conservation of baryon number as an *empirical* law. But combining quarks and leptons into a single symmetry banished conservation of baryon and lepton numbers. Consequently, grand unification *allows* a proton to decay into a lepton, and if something is *possible* in the quantum realm it *will* occur sooner or later. *Baryonic matter* is therefore inherently radioactive and will eventually decay.

In terms of quarks, a proton comprises a 'd' and a pair of 'u' quarks. The decay could proceed by one of several possible interactions. For example, if one of the 'u' quarks were to emit an X-boson that interacted with the 'd' quark, this would turn the 'd' quark into an antielectron and the 'u' into an antiquark, which would then be obliged to pair with the remaining 'u' as a neutral pion that would eventually decay into a pair of gamma rays.

The Pati–Salam SU(4) theory did not offer a value for the proton's lifetime, but Georgi, Quinn and Weinberg calculated that SU(5) implied a lifetime of between 10^{31} and 10^{32} years. In light of the fact that the Universe is only 10^{10} years old and that considerable baryonic matter is still present, it might appear that proton decay is

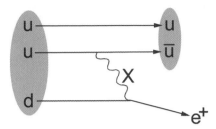

A proton comprises two 'u' quarks and a 'd' quark. One scenario for proton decay involves one of the 'u' emitting an X-boson and transforming itself into the corresponding antiquark, with the boson interacting with the 'd', turning it into an antielectron. As the colour-neutral antielectron escapes, the 'u' and its antiparticle pair up as a neutral pion. Experiments to infer this process involved surrounding a tank of exceedingly pure water with photomultipliers to detect the pair of photons from the subsequent decay of the pion.

of little significance, but this is not so because *lifetime* is of a statistical nature and the process is ongoing. Maurice Goldhaber pointed out that the proton's lifetime must be considerably longer than the age of the Universe; if it were not then the ongoing rate of decay would make our own bodies highly radioactive. As he put it, we would "feel it in our bones". The greatest possible radioactivity that our bodies could yield indicated that the proton's lifetime could not be less than 10^{16} years, but this barely moved the lower limit towards the prediction. In 1953 Goldhaber had worked with Frederick Reines and C.L. Cowan to verify the existence of the neutrino using a tank of water, and he now realised that this data also gave a lower limit for the lifetime of the proton at 10^{22} years. A more modern neutrino detector installed three kilometres down a South African gold mine, and operated by the University of Witwatersand in collaboration with the University of California at Irvine and Case Western Reserve University, would also have noted any proton decays, but in 1974 it was announced that after three years of operation it had detected nothing relevant, and this pushed the lower limit to 10^{30} years. The fact that this was marginally less than the predicted range prompted a race to be the first to confirm proton decay, not only as a test of the SU(5) form of grand unification but also in order to infer the ultimate fate of baryonic matter, which was no trivial issue. Given that proton decay was rare, the only way to confirm it was to observe an extremely large number of protons and spot the few instances occurring over a reasonable operating interval – several months, at least.

Strategies were developed using 'dense' and 'sparse' detectors. The first claim to have found proton decay was made by a collaboration between India and Japan utilising a 150-tonne block of iron (hence the term 'dense') emplaced two kilometres down the Kolar gold mine in India. This, however, was received with scepticism because an unambiguous identification of the decay products in a solid mass is much more difficult than in a cloud or a bubble chamber. A sparse detector was a vast tank of extremely pure water placed deep underground to shield it from cosmic rays and other sources of 'events' which might mask the signature of any decaying protons. The tank was then monitored for the tell-tale double gamma rays. In Europe such detectors could be readily placed in Alpine tunnels, and one was put in a service cavern under Mont Blanc; in North America, however, as there were no such convenient locations, deep mines were used. The first detector to enter service, in 1981, was in an iron mine north of Minneapolis by the University of Minnesota. A consortium of the University of California at Irvine, the University of Michigan and the Brookhaven National Laboratory (and therefore known as IMB) commissioned a detector of 10,000 tonnes of highly pure water in a Morton–Thiokol salt mine near Cleveland in Ohio (in fact, it was 2,000 feet beneath Lake Erie) in late 1982. The first three months of the IMB operation produced only one candidate for proton decay, implying a lower limit of 6×10^{31} years. After three years this had been pushed to 10^{33} years. At that point, however, the background 'noise' from neutrinos swamped the 'signal'. In 1996 a Japanese–American consortium commissioned the 'Super-Kamiokande' neutrino detector down a deep zinc mine near Tokyo, and this pushed the lower limit towards 10^{34} years. Although SU(5) could be ruled out on this basis, the search for proton decay continues in the hope of eventually verifying that the proton is unstable, but pushing the limit rapidly becomes an increasingly time-consuming process.

Although regarded as a great step forward, SU(5) was rejected as a unification of the fermions. Nevertheless, as A.H. Guth reflected, it was "very impressive that one simple theory could incorporate so much physics". In a sense, physicists had been dreading its confirmation because if it were true then it meant that there would be *no new physics* to be discovered between the energies corresponding to the spontaneous breaking of the grand unified symmetry and the subsequent loss of the electroweak symmetry. Consequently, with contemporary accelerators at 1 TeV, there was little prospect of ever crossing what was dubbed the *asymptotic desert* to 10^{15} GeV.

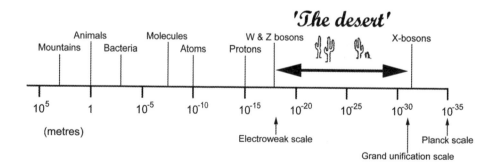

Physicists were so appalled at the SU(5) theory's implication that no new physics would be discovered at energies above that at which the electroweak symmetry was broken at 100 GeV until the unification of fermions at 10^{15} GeV, that they dubbed this barren span the 'asymptotic desert'.

THREE GENERATIONS

By eliminating the distinction between quarks and leptons, the grand unification theory required a balance in their numbers. Even as the discovery of the charmed quark in November 1974 established four of each, Martin Perl was accumulating SPEAR data that hinted at a heavy relative of the electron and muon – in fact, it had almost twice the mass of a proton! By the time of the formal announcement of this 'tau lepton' in 1976, physicists were sufficiently confident to predict that, in addition to a matching neutrino there ought to be a third pair of quarks. Although, in 1977, L.M. Lederman at Fermilab identified the 'bottom' quark in the form of the upsilon meson, its 'top' counterpart was not spotted until 1995. The tau-neutrino was tracked down in 2000. Would quarks continue to turn up, or had nature imposed a limit on this series of 'generations'? In 1964 V.L. Fitch and J.W Cronin showed that CP symmetry was violated by kaon decay, and in 1972 Makato Kobayashi and Toshihide Maskawa in Japan showed that *three or more* generations of quarks were required to explain this violation. In 1976 D.N. Schramm, J.E. Gunn and Gary Steigman inferred from the primordial nucleosynthesis of helium that there can be *at most four* neutrino types, and recent particle experiments imply that there are *only three*, so in this respect the end should be in sight for the particle hunters.

-1	$-\dfrac{1}{3}$	0	$+\dfrac{2}{3}$
e	d	ν_e	u
μ	s	ν_μ	c
τ	b	ν_τ	t

It seems highly likely that not only is there a balance in the numbers of quarks and leptons, but there are three generations of each. It remains to be determined *why* this is so. The matter with which we are familiar can be explained solely in terms of the first generation. The exotica of particle physics are associated with higher energies. The numbers give the electric charges of each column.

SUPERSYMMETRY OF FERMIONS AND BOSONS

Although it appeared that nature had not used a specific symmetry to unify the fermions, perhaps the individual symmetries of the Standard Model had arisen from the spontaneous breaking of an even greater symmetry that unified the fermions and bosons. Such a *supersymmetry* was developed by Julius Wess and Bruno Zumino of CERN in 1973. They did not provide a full theory, merely a formalism for exploring the issues of combining particles embodying *differing spins*. In quantum field theory, intrinsic spin is a geometric manifestation of rotational symmetry: that is, although a boson is returned to its initial state by a 360-degree rotation, a fermion does so only after a 720-degree rotation. If transformations between these geometries could be defined, then fermions and bosons would be equivalent to 'projections' of a single underlying geometric entity. In this multi-dimensional structure, the four space-time dimensions are part of a 'superspace', and each of the known particles has a supersymmetric partner, such that the partner of a spin $= 1/2$ fermion is a spin $= 0$ boson and that of a spin $= 1$ boson is a spin $= 1/2$ fermion.* This scheme embodied an exact symmetry, but the fact that the known fermions and bosons do not match up implies that this symmetry is broken and that the partners are exceedingly massive. For the experimental physicists, therefore, supersymmetric partners were as frustrating as the X-bosons of the grand unified theory of fermions.

SUPERGRAVITY

Even if fermions and bosons *could* be unified by a supersymmetric quantum field theory, it would *not* incorporate gravity. Perhaps we should try to incorporate both gravity and the strong force with electroweak in a *single* step. This strategy followed Albert Einstein's dream of developing a 'unified field theory' that incorporated all of the forces. Shortly before Einstein's death in 1955, L.V. de Broglie pointed out that, "Theoretical physicists are at present divided into two apparently irreconcilable groups. On the one hand, Einstein and his followers are trying to develop general relativity theory, while by far the great majority of theorists, attracted by atomic problems, are trying to develop quantum physics quite independently of general relativity", with the two groups "utterly ignoring each other". In 1962 J.A. Wheeler at Princeton envisaged black holes as the meeting ground between general relativity and quantum mechanics. In developing a superspace, he realised that there was a unit of *length* of 10^{-35} metre (known as the Planck–Wheeler length, often abbreviated to the Planck length). If a quantum was smaller than this, it would behave like a black hole: its gravitational field, while minuscule, would nevertheless be sufficient to warp space to cut itself off from the observable Universe. Wheeler said that since nothing smaller than this could be *observed*, space itself was *quantised*, and that on this scale geometry, space and time all lost meaning. In effect, he posited, space was *made* of black holes. This was a revolutionary idea at odds with general relativity's view of space in terms of a continuum.

---

* In the nomenclature dubbed 'slanguage' by Murray Gell-Mann, the bosonic quarks, electrons and neutrinos became 'squarks', 'selectrons' and 'sneutrinos', and the fermionic photons, W, Z and gluons became 'photinos', 'winos', 'zinos' and 'gluinos'.

The first *supergravity* was devised in 1976 by Sergio Ferrara in Rome and D.Z. Freedman and Peter van Nieuwenhuizen of the State University of New York, but it was soon superseded by a simpler theory developed by Bruce Zumino and Stanley Deser. Supergravity was a supersymmetric extension of general relativity in which the superpartner of the spin = 2 graviton was the spin = 3/2 'gravitino'. In fact, a class of theories were identified (labelled $N = 1$ to $N = 8$) differing in the number of particles. The simplest ($N = 1$) had only the graviton–gravitino. Although it clearly did not have sufficient richness to account for reality, the fact that the infinities introduced by the graviton and gravitino cancelled one another out raised the prospect of developing a 'quantum gravity'. However, it was not evident that the more general versions of the theory would be renormalisable. The most complex version ($N = 8$) was sufficiently rich to account for reality, but none of its fields could be identified with the familiar particles.

Another approach to incorporating gravity was to explore 'higher' dimensions. In 1919 Theodor Kaluza, a mathematician at the University of Königsberg, extended the space-time of general relativity by introducing a *fifth* dimension, and was amazed to discover that this 'degree of freedom' obeyed Maxwell's equations, implying that electromagnetism, like gravitation, is a manifestation of geometry. To account for the absence of a macroscopic manifestation of this fifth dimension, Kaluza proposed that in addition to the known 'extended' spatial dimensions there are others that are 'curled up' – just as a cartesian axis system can be thought of as passing through each point in three-dimensional space and extending to infinity, this extra dimension can be imagined as being a tiny circle which, although infinite, is bounded. In 1926 Oskar Klein in Sweden generalised the theory to arbitrary dimensionality, recast it in terms of quantum mechanics, and calculated the radius of the curled up dimension as being no greater than 10^{-32} metre. Unfortunately, this scale was so far below observability that the idea was untestable. In 1978, however, Eugene Cremmer, Bernhard Julia and Joël Scherk in Paris discovered that combining Kaluza–Klein theory with supergravity resulted in an 11-dimensional space-time, but after several years this was established to be flawed, and supergravity once again fell by the wayside.

STRING THEORY

Even as grand unification of fermions, supersymmetry of fermions and bosons, and supergravity theories were being investigated and discarded, work on a radically different idea was progressing.

In 1968 Gabrielle Veneziano, a theorist at CERN, found that for some reason the 'beta function' devised by Leonard Euler two centuries ago as a mathematical exotica was able to express aspects of the strong force that acted between hadrons. In 1970 Yoichiro Nambu of the University of Chicago, Holger Neilson of the Bohr Institute in Copenhagen, and Leonard Susskind of Stanford University explored the physical interpretation of Euler's function and discovered that it modelled point-like particles as vibrating one-dimensional 'strings'. The infinities of quantum field theory are due to its viewing particles as dimensionless points that can be brought arbitrarily close together and, by implication, create infinitely strong forces. A string does not invite this dilemma. A string can be 'open' with its ends flapping, or 'closed' into a loop. An open string can mimic a spin $= 1$ boson, so the initial focus was on 'open' strings. A string is not *something in space*, it is a piece *of space*. In contrast to a violin string whose oscillation is damped out as it transfers its energy to its environment, a string of space is isolated, and because its energy is contained its oscillation is undamped. Since they can vibrate in an infinite number of resonant 'modes', such strings can mimic an infinite number of types of particle, each of which has a specific energy. Although the theory could be adapted to cope with the realisation that hadrons are clusters of quarks, the fact that it predicted so many more types of particles was regarded as a serious flaw. Nevertheless, the idea of strings was intriguing, and when generalised in 1971 by Pierre Ramond at the University of Florida to include fermions as well as bosons it was found to constitute a new type of relativistic quantum theory. Furthermore, because patterns of string vibrations arise in pairs, differing by a half-unit of spin, it is *intrinsically supersymmetric*. Although it is more properly 'superstring' theory, most people elide the prefix.

In 1974 J.H. Schwarz at Caltech and Joël Scherk in Paris independently realised that one of the closed string modes had properties resembling the spin $= 2$ graviton, which raised the possibility of describing gravitation in terms of a quantum theory *without directly involving fields*. However, as researchers focused on uniting quantum forces by developing ever broader symmetry groups, this suggestion fell upon deaf ears. In 1984, Schwarz, with M.B. Green at Queen Mary College in London, refined the string formalism further, and postulated that it was capable of uniting *all* four fundamental forces. To be internally consistent, the theory *required* curled up spatial dimensions. Intriguingly, string theory has *only one free parameter*, which is the 'tension' in the string. In the initial version of the theory, this had been scaled to match the energies of the strong force, but the identification of the graviton enabled Schwarz to calibrate the tension. The *minimum* energy of a string is *precisely* the 'Planck energy'. As this is multiplied by the number of wavelengths in the string's mode, a string can have an *exceedingly large* energy. However, quantum fluctuations lead to cancellations, and so strings *manifest themselves* at much lower energies corresponding to the *residuals*. If the cancellation is exact, then such a string appears to be massless. As for the other fundamental charges, spin is simply an aspect of the manner in which a string vibrates. As vibrations in strings can extend over curled up dimensions, it is easy to visualise a fermion slowly 'precessing' in another dimension, and thereby requiring a 720-degree rotation in the familiar three-dimensional space to resume its initial state. If mass and spin are *manifestations of geometry*, might not the 'charges' of the various forces be too – as Kaluza had suspected for electromagnetism. The prospect of a 'theory of everything' was dizzying. A flurry of activity over the next few years established that string theory *required* many of the features that the Standard Model left to be empirically derived. However, by 1986 there were *five* versions of the theory, each pairing its fermions and bosons differently, and this was seen as a serious problem. There was a hiatus as the detail was refined, and then in 1995 Edward Witten of the Institute for Advanced Study in

Princeton launched a second frenzy of activity that unified the five versions of the theory in the context of an 11-dimensional space-time – as hinted at by the supergravity studies. This was promptly dubbed 'M-theory', but there would appear to be as many meanings for 'M' as there are researchers. It encompasses not only the familiar one-dimensional strings, but also 'membranes' in arbitrary dimensions (prompting the alternative name of 'brane' theory). One early discovery was that whereas in the earlier theories the gravitational coupling constant did not match that of the quantum forces until an energy density slightly higher than the unification energy, in M-theory it merged with them *at precisely* that energy.

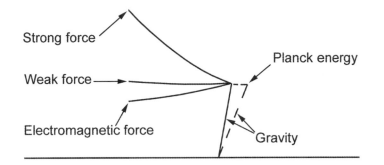

A welcome prediction of M-theory is that gravity unites with the non-gravitating forces at a single energy (solid line); in earlier theories (dashed line) this was not so.

Lest the idea of a one-dimensional string sound ludicrous, it is no more so than a dimensionless point. A string is so small that its dimensionality is apparent only at *extremely short range*. In fact, at about 10^{-35} metre, a string is 10^{-20} of the diameter of a proton. It will resemble a point even in deeply penetrating interactions. On this scale, gravitation rises in strength to rival that of the other forces. A physical theory dealing with this Planck scale *must* therefore unify the quantum forces with gravity. However, as noted, efforts to develop quantum field theory into supergravity theory failed.

The incompatibility between general relativity and quantum field theory derives directly from the Uncertainty Principle which forms the basis of quantum theory. In terms of general relativity, gravitation represents the manner in which the presence of mass (or more accurately mass-energy, because mass is just concentrated energy) curves space. In the absence of mass, space must necessarily be 'flat' and smooth. Consequently, in the absence of a significant mass (such as in the vacuum of space) the gravitational field should be zero. However, thanks to the Uncertainty Principle, the vacuum of the quantum realm is seething with energy. Whereas general relativity requires the gravitational field on the scale of the quantum realm to *be* zero, the Uncertainty Principle requires only that it maintain *an average* of zero, and because gravitation is a measure of the curvature of space and its value fluctuates wildly, space, far from being smooth, is so warped as to constitute a random froth – using J.A. Wheeler's term, it is "foamy". As a result, the equations of general relativity produce infinites. These derive from the fact that quantum field theory deals with fields originating from point sources. It is apparent that neither quantum field theory nor general relativity is a complete description of reality; each is actually only *an approximation* in its own realm to something deeper that applies at all energies up to the Planck energy. With one-dimensional strings, such infinities do not arise. The energy of a resonating string depends on the amplitude of the oscillation and its frequency (that is, the number of whole wavelengths contained within its length, or in the case of a loop, within its circumference) and more frantic vibrational modes have greater energy than tranquil ones. A calculation revealed that the 'charges' of the fundamental forces (that is, electric charge, colour, flavour, etc.) derive from aspects of the modes in which strings vibrate. Furthermore, the theory describes not only the fermions on which forces act, but also the mediating bosons. Quantum field theory has a *distinct* field for each fundamental charge, and particles are bundles of energy corresponding to oscillations in these fields. In string theory, all charges are aspects of the manner in which a string vibrates. The theory does not say that a particle that quantum field theory would consider to be point-like 'contains' a string, it *is* a string. A snapshot at a particular time will show a multitude of strings, but a 'long exposure' would record their loops tracing out tubes. This provided an entirely new prespective on particle interactions. In fact, if we 'stand back', particle interactions are seen to weave together into a *static* structure in space-time.

In string theory, particles are tiny loops. When they interact, two loops merge and split apart again in a different configuration (left). If time is taken into account by envisaging the moving loops as drawing out 'tubes' then particles and their interactions can be thought of as a static network of interconnecting tubes in space-time (right).

Just as Newton's laws of motion are a low-speed approximation to Einstein's relativity, quantum field theory is a low-energy approximation to string theory. In terms of the quantum realm, gravitation 'unifies' only at high energies. The fact that string theory encompasses *both* gravitation and quantum field theory suggests that it will not be possible to push quantum field theory beyond the Standard Model, and we should employ string theory to study the *much higher* energies on the far side of the 'asymptotic desert'. Indeed, during the decade in which string theory was consolidated, there was no progress whatsoever in extending the Standard Model by the conventional approach other than to assert that it is *probably* supersymmetric. There is circumstantial evidence for this. In the original formulation of the Standard Model, terms in the equations are 'fine tuned' to yield sensible results. This alarming sensitivity is absent in supersymmetry, because the irksome effects of fermions are cancelled by those of their bosonic counterparts, as are those of bosons by their fermionic counterparts. Consequently, even though we are not be able to detect the supersymmetric particles directly, their presence in the seething vacuum ought to be significant. In 1991 Ugo Amaldi, Wim de Boer and Hermann Fürstenau at CERN measured how the coupling constant of the weak interaction varied at high energy in order to recalculate the unification energy (in effect, revising the figure estimated in 1974 by Georgi, Quinn and Weinberg) and discovered that while the strengths of the quantum forces converged they did not actually intersect, and that this discrepancy was eliminated in a *Supersymmetric Standard Model*.

As yet, however, there is no test to *decisively* indicate that strings lie at the heart of what we had presumed to be point-like particles – the energies needed to establish this lie far beyond the capabilities of our accelerator technology. Nevertheless, one 'prediction' of string theory is gravitation. While this may sound trivial, Newton and Einstein devised theories in order to explain observations but string theory *requires* gravity to exist, and with its observed characteristics. In the same sense that general relativity was intrinsically better by requiring the inverse square law which Newton had incorporated empirically, string theory is better still. Furthermore, string theory unites gravitation with the consequences of the Uncertainty Principle on the Planck scale. String theory subsumes both general relativity and quantum field theory. Also, whereas the Standard Model does not require supersymmetry, string theory *does*. The Standard Model is silent on why there are three families of quarks and leptons of successively greater masses, but a relationship between the curled up dimensions of string theory offers a hint of why this should be the case. In his early enthusiasm for quantum field theory, Freeman Dyson speculated that it might describe 'reality', but it is only an approximation; perhaps we might reasonably claim this of string theory.

Throughout most of the development of physics, experimenters have pioneered and theorists have sought to make sense of their observations. From time to time, for brief periods, theorists forged bold predictions which experimenters raced to verify. However, the energies involved in investigating the fabric of space-time are so beyond the capability of our technology that the theorists are now essentially on their own.

Part III

Discovering the Universe

7

The spiral nebulae

WHIRLPOOLS IN THE SKY

In the grand tradition of the financially independent 'gentleman astronomers' of the mid-nineteenth century, William Parsons decided to construct the most powerful telescope in the world. His aim was to determine the nature of the nebulae. Since Herschel's time, most astronomers had switched to the new achromatic refractors, but the lenses had narrow apertures and Parsons required 'light grasp'. After making a number of small mirrors in order to develop his innovative manufacturing technique, in 1840 he made a 36-inch mirror whose quality was so good that it prompted him to start work on a 72-inch as his primary instrument. In 1841 he inherited the title of the third Earl of Rosse and took up permanent residence in Birr Castle in Ireland. In 1845 the 72-inch mirror was set in a 56-foot-long wooden tube, and put in a massive stonework transit-style mounting whose range of motion was limited to 15 degrees to either side of the meridian. He had therefore to wait until an object neared the meridian to observe it. The first target for the 'Leviathan' (as the massive telescope became known) was M42 in the Sword of Orion, whose nebulosity Parsons found to be 'streaky'. Delighted with the optical quality of his telescope, he made a five-year survey of the heavens. Upon finding M51 in Canes Venatici to possess a spiral pattern, he named it the Whirlpool nebula. His discovery of a total of 15 such spirals was his primary contribution to astronomy.

In 1755 Immanuel Kant proposed that the Sun and its retinue of planets formed when a cloud of gas collapsed under its own gravitational attraction, and in 1796 the French mathematician P.S. Laplace investigated how this might have occurred. In his *Essay on the System of the World* he stated that since the Sun's rate of rotation would have increased as it contracted, it would have progressively shed angular momentum by ejecting a succession of co-planar rings which, he argued, went on to condense to form the planets. The spirals discovered by Parson were widely presumed to be 'Laplacian nebulae' in which planetary systems were forming around newborn stars.

In 1845 William Parsons (the third Earl of Rosse) commissioned the most powerful telescope in the world at his residence in Birr Castle in Ireland. Although its 72-inch mirror gave unprecedented 'light grasp', the transit-style mounting limited its range of motion to 15 degrees to each side of the meridian line.

THE MILKY WAY SYSTEM

In 1815, Harvard College began to consider the construction of an observatory, and W.C. Bond was sent to Europe to inspect the workings of similar institutions. Upon his return, he submitted his report and promptly established his own observatory at Dorchester in order to test techniques. When the Harvard College Observatory was opened in 1844 with a fine 15-inch refractor supplied by Merz-Mahler of Germany, Bond was appointed as director.

The first photographic process was invented by Frenchman, L.J.M. Daguerre, in 1830, but it required very long exposures and its astronomical utility was limited. Nevertheless, in 1840 J.W. Draper secured a photograph of the Moon using a simple lens to focus its image. In 1849 W.C. Bond took a telescopic picture of the Moon, which his son, G.P. Bond, showed at the Great Exhibition in London in 1851, where it caused a sensation. In 1850 W.C. Bond and J.A. Whipple captured an image of Vega. In 1851 F.S. Archer in England invented the 'wet' collodion process for glass-coated plates, and as this required much shorter exposures than Daguerrotype, Warren de la Rue in England began taking telescopic collodion pictures in 1852. In 1856 N.R. Pogson in England formalised Hipparchus's scale for the brightness of stars by introducing a mathematical relationship in which a difference of five magnitudes corresponded to a 100-fold difference in brightness (on this scale, the Sun is magnitude −27). In 1871 R.L. Maddox in England introduced 'dry' plates that could undertake much longer exposures and thereby capture much fainter stars.

When William Parsons (Lord Rosse) viewed M51 through his 'Leviathan' the distinctive spiral structure prompted him to name it the Whirlpool nebula. His sketch (top) compares well with the modern view, as represented here by an electronic image taken using the prime focus camera on the William Herschel Telescope in June 2000 by Javier Méndez of the Isaac Newton Group of Telescopes at La Palma and Nik Szymanek of the Society for Popular Astronomy and a member of the amateur UK Deep Sky CCD imaging team.

The Orion Nebula (M42, NGC1976) surrounding the star theta Orionis in Orion's Sword, documented in context by the Two-Micron All Sky Survey. Courtesy of the University of Massachusetts and the Infrared Processing and Analysis Center at JPL/ Caltech.

The ionised hydrogen of the Orion Nebula is energised by the very hot young stars in the Trapezium cluster, embedded in the brightest part of the cloud. This picture was taken in 1973 using the 4-metre telescope at Kitt Peak National Observatory, near Tucson, Arizona. Courtesy of Bill Schoening/NOAO/AURA/NSF.

In the 1840s and 1850s, F.W.A. Argelander of the Bonn Observatory had utilised a 3.4-inch Fraunhofer refractor to chart the entire northern sky down to ninth magnitude, and between 1859 and 1862 he published the Bonn Survey (this *Bonner Durchmusterung* is known by its 'BD' star designations). With 324,198 stars, the focus was on quantitative data. Argelander also introduced a nomenclature for variable stars – a scheme that is still in use, although in modified form. Upon being appointed as director of the Cape Observatory in 1879, Scotsman David Gill set out in 1882 to produce a photographic map of the southern sky. As the analysis of the plates occupied the Dutch astronomer J.C. Kapteyn for a decade, it was 1904 before the Cape Photographic Durchmusterung was issued.

The French philosopher Auguste Comte lamented in 1835 that the constitution of the stars would "forever remain unknown". After G.R. Kirchhoff and R.W. Bunsen of the University of Heidelberg had interpreted spectral lines in 1859, William Huggins in London, upon hearing that spectroscopy could reveal the chemical composition of the stars, metaphorically compared it to "coming upon a spring of water in a dry and dirty land". He had a telescope with an 8-inch lens, made by Alvan Clark of Massachusetts, which he fitted with a spectroscope. He delightedly wrote in his diary that nearly every night's work was "red-lettered by some discovery". Although the spectrum of even the brightest of stars was so dim that it was difficult to discern any detail, in 1863 he was able to announce that they were continua similar to the Sun's. While conducting his sweeps of the heavens, William Herschel had seen a number of roundish nebulae. As each of these 'planetary nebulae' (as he named them) had a single faint star at its heart, he concluded that the nebulosity could not be unresolved stars and decided that it must be 'luminous fluid'. In 1864 Huggins inspected the spectrum of a bright planetary nebula and, to his amazement, found only a single glowing line of an incandescent gas. It was the same for M42 in Orion and M1 in Taurus (which Parsons had named the Crab on account of its shape). By 1866, he had inspected 60 nebulae and found that while one-third were gaseous, the rest, including Parsons's spirals, had continuous spectra, which implied that they were not Laplacian nebulae, but systems of stars whose members could not be individually resolved.

C.J. Doppler in Austria realised in 1842 that the wavelength of light should vary in proportion to the relative speed of its source, but it was 1848 before the details were refined by A.H.L. Fizeau in France. Initially dubbed the Doppler–Fizeau effect, this is now known simply as the Doppler effect. In 1868 Huggins used the displacement of a hydrogen line to measure the radial velocity of Sirius, which he found was receding at 50 kilometres per second. He painstakingly compiled radial velocities for 30 stars by visual observation. In 1872 Henry Draper succeeded in photographing four lines in Vega's spectrum with a 28-inch reflector. H.C. Vogel at the Potsdam Observatory took up stellar photography in 1873, and by 1888 was routinely cataloguing radial velocities. J.E. Keeler of the Allegheny Observatory in Pittsburg was also measuring radial velocities by 1890. In an *Atlas of Stellar Spectra*, published in 1899, Huggins reviewed everything that was known on the subject. In 1882, the year that Henry Draper died, E.C. Pickering, who had been appointed director of the Harvard College Observatory in 1876, built an 'objective prism' in order to photograph a spectrum for each star in the field of view simultaneously, instead of doing so one by one. The resolution was low, but sufficient to establish general characteristics. Pickering ordered a photographic campaign to compile a star catalogue. In 1891 he dispatched his younger brother, W.H. Pickering, to establish an observatory at Arequipa in the Peruvian Andes in order to extend this survey to the southern hemisphere. The first classification of stellar spectra was proposed in 1867 by P.A. Secchi in Italy, on the basis of 4,000 examples. In 1868, after a decade of study, A.J. Ångström in Sweden published a chart of 800 lines in the solar spectrum. After comparing the spectra left by Draper to those of the Sun, Pickering introduced a new classification scheme, for which A.C. Maury and A.J. Cannon, two of his many female assistants, invented a nomenclature using letters and numerical subdivisions. Upon joining Pickering's team in 1895, H.S. Leavitt devised a photometric technique for measuring the brightness of stars on photographic plates. Upon becoming a full member of staff in 1902, Leavitt was appointed head of the photometry department. After many years of processing, the Henry Draper Catalogue (known by its HD star designations and the HDE southern 'extension') was finally released in 1924 with more than 250,000 positions, motions, brightnesses and spectral classifications.

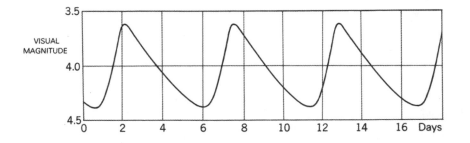

The light curve of the star delta Cephei varies in a regular manner between well-defined maxima and mimima over a 5.4-day period.

John Goodricke in England was a deaf-mute amateur astronomer who specialised in variable stars. In 1878, at age 18, he inferred from the 'light curve' of Algol (beta Persei) that a faint companion was periodically passing in front of the primary star in a binary system, partially obscuring it. Large numbers of these 'eclipsing binaries' were later catalogued. In 1784, shortly before his death, he realised that delta Cephei fluctuates in a regular manner over a 5.4-day period, rapidly rising to a given peak magnitude and then slowly fading back to the same minimum. Over the years, other 'cepheids' (as they came to be known) were found with periods ranging from about a day to several months, the mean being about a week. In 1894 A.A. Belopolski at the Pulkova Observatory noticed a radial velocity cycle in the spectrum of delta Cephei synchronised with its light curve. The spectral type also oscillated, implying a variation in temperature. In 1908 S.I. Bailey at Harvard found variable stars in globular clusters. In addition to a few cepheids with periods of 12 to 20 days, there

The businessmen Andrew Carnegie funded G.E. Hale's commissioning in December 1908 of the Mount Wilson Observatory's 60-inch reflector. Although its collecting area was smaller than Parsons's Leviathan, the 60-inch mirror was more efficient and therefore had greater effective light grasp. Furthermore, it was a much more capable instrument because the mountaintop site had better seeing and the telescope was on a fully-steerable mount.

were many RR Lyrae variables which were similar, but fainter, and ran through their cycles in about 12 hours. Such stars would eventually play a key role in discovering the true scale of the Universe.

Having attended a general course on astronomy at the University of Missouri in 1907, Harlow Shapley gave up his plan to train as a journalist and pursued this new interest instead. He received his doctorate from Princeton University in 1914 for an investigation of eclipsing binaries under the supervision of H.N. Russell. As Shapley was planning his post-doctoral work, G.E. Hale, the director of the Mount Wilson Observatory, invited him to use the 60-inch reflector, which was the most powerful telescope in the world. Shapley accepted. Upon being appointed director of the Cape Observatory in 1834, John Herschel had erected his father's favourite 20-foot reflector and made a comprehensive survey of the southern sky. On returning to London in 1838 he ceased observing and launched himself into the great task of processing his accumulated observations, and in 1847 he issued a catalogue which included a list of 2,000 nebulae, increased to almost 100 the list of globular clusters, and showed that most of them were located in the constellation of Sagittarius. Shapley decided to discover why this was the case. Mindful of the findings of Belopolski and Bailey, he focused on the RR Lyrae variables (which he referred to as 'cluster variables'). They varied so rapidly that when Bailey had taken long exposures in order to follow them as they faded to minima, the loss of temporal resolution had made it difficult to trace their light curves. Using the 60-inch's light grasp, Shapley was able to monitor their fluctuations in detail. Furthermore, he recorded their spectra and observed that their spectral lines were shifted blueward while they were at their brightest, and redward at their faintest. Such oscillations meant that the stars were *pulsating*, with their radius inflating and contracting by as much as 10 per cent in only a few hours.

By 1900, some 65 stellar distances had been determined, essentially by the same method used by Bessel. Direct parallax measurement was made simpler and more accurate by the introduction of astrophotography, but by the turn of the century it had become evident that indirect methods would have to be developed. Once both proper motions and radial velocities became available for large numbers of stars, it was possible to identify their true motions through space. As William Herschel had realised, there was a general drift in which widely separated stars shared a common motion. Not all stars had *precisely* the same vector, of course. The degree to which a star's true velocity differed from the mean could be translated into a unique motion across the sky, and the remainder could be interpreted as a parallax measure. As this was a statistical process, this method is referred to as 'statistical parallax'. Setting the basis for another method that involved star clusters, R.A. Proctor had pointed out in 1869 that five of the stars in the 'Plough' of Ursa Major had parallel proper motions, and so were moving through space as a group. As proper motion catalogues were extended to fainter stars, it became evident that there were many of these 'open' clusters. In 1908 Lewis Boss, the director of the Dudley Observatory in Albany, New York, made a study of the Hyades cluster in Taurus. The manner in which the proper motions converged enabled the cluster's motion in space to be determined, so that the residual proper motions of the individual stars could be interpreted as the angle for a parallax measure. The application of these methods revealed that the luminosity of a star was related to its spectral class. This meant that once the distance of *one* star of a given class had been determined and its luminosity calculated, the distances to similar stars could be inferred from apparent brightness. After working as a chemical engineer in St Petersburg, Ejnar Hertzsprung took up amateur astronomy and in 1909 became a lecturer in astrophysics at the University of Göttingen. In 1913 Hertzsprung and H.N. Russell independently plotted stellar luminosity versus spectral class (and therefore temperature). The distribution of this Hertzsprung–Russell diagram was not random – there was a strong concentration along the diagonal from luminous-and-hot to cool-and-faint, a line dubbed the 'main sequence'. It was soon noticed that the plots for clusters 'departed' from this line at different points, with the older-looking clusters doing so at cooler temperatures. Hertzsprung realised that there were *two* types of red star – extremely luminous 'red giants' and very dim 'red dwarfs' – and he gleaned from this an early insight into stellar evolution.

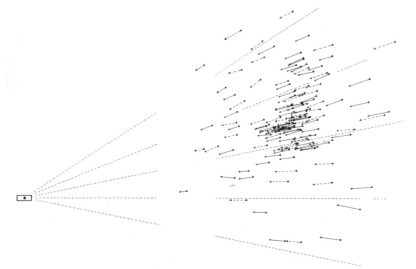

In plotting the proper motions of the stars in the Hyades cluster in Taurus in 1908, Lewis Boss noted an apparent convergence, indicating that the cluster as a whole is receding from us at an oblique angle. The departures from the mean are the result of the motions of the individual stars within the cluster.

In 1924 A.S. Eddington at Cambridge noted from a study of binary systems that the luminosity of a star was proportional to its mass. This discovery explained the main sequence. In 1926 Eddington published a book entitled *The Internal Structure of the Stars*, which explained, for the first time, that a star could be modelled as a large ball of gaseous hydrogen with a core sufficiently hot to drive nuclear fusion reactions (the solar photosphere is 6,000 K, but its core is several millions of degrees hotter). After graduating from the University of Frankfurt and gaining his doctorate in Munich in 1928, H.A. Bethe fled to England in 1933 and moved on to America in

The southern sky, set against the dome of the 4-metre Blanco Telescope at the Cerro Tololo Interamerican Observatory in Chile, was photographed by Roger Smith. In addition to a section of the Milky Way, the Large and Small Magellanic Clouds are visible on the left. Courtesy of NOAO/AURA/NSF.

1935 in order to take a post at Cornell University. In 1938 Bethe worked out the details of the 'hydrogen burning' processes that convert mass directly into energy in stellar cores. As it begins to burn the products of earlier fusion (starting with helium) the star 'evolves off' the main sequence. One outcome of this insight was that it was possible to estimate the age of a globular cluster from the point where it departed the main sequence.

When Ferdinand Magellan sailed into the far southern ocean in 1519, he noted in his log the presence of two bright patches in the sky that looked just as if they were fragments of the Milky Way that had somehow drifted loose. Although his ship, *Victory*, completed the historic circumnavigation of the globe in 1522, Magellan had died *en route*, and so the two celestial clouds were named in his honour.

A series of photographs of the Magellanic Clouds had been taken by the 10-inch refractor at Arequipa in order to identify variable stars. Leavitt started the analysis of these plates in 1904, and by 1908 she had recorded 25 cepheids in the Small Magellanic Cloud (SMC). She noticed that those with longer periods appeared brighter. By 1912 she had a comprehensive archive, and upon plotting the periods against apparent maximal and minimal brightnesses, she confirmed her earlier impression and derived a smooth curve. As the stars were all at essentially the same distance, this relationship applied to *intrinsic luminosity* (although only in a relative

The Small Magellanic Cloud (SMC) is a nearby irregular dwarf galaxy. This picture was taken using the Curtis Schmidt astrograph at the Cerro Tololo Interamerican Observatory in Chile. Courtesy of NOAO/AURA/NSF.

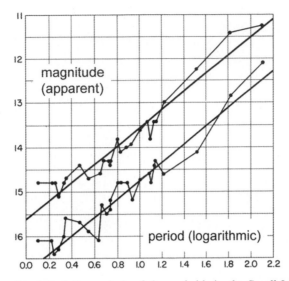

When H.S. Leavitt plotted the periods of the cepheids in the Small Magellanic Cloud against their maximal and minimal apparent brightnesses, she realised that there was a simple relationship between the period and the luminosity of such stars. (*Periods of 25 variable stars in the Small Magellanic Cloud*, Harvard College Observatory Circular no. 173, 1912.)

sense, because the actual distance was unknown). Although Leavitt realised the significance of this discovery, and was eager to pursue it, Pickering insisted that she continue her work towards the Henry Draper catalogue.

Leavitt's plot relating the periods and luminosities of the cepheids was calibrated in 1913 by Hertzsprung. Since no cepheid had a measurable parallax, and there were none in the open clusters, he utilised a variation of the statistical parallax method to estimate the distances to 'nearby' cepheids, thereby deriving the luminosities of 13 cepheids with periods ranging from 1.3 to 66 days. In 1906 Kapteyn had reported a photographic survey replicating Herschel's process of counting stars in selected areas. This time the 65 known parallaxes gave a sense of depth. Like Herschel, Kapteyn concluded that the Sun was located near the centre of a disk of stars, which he estimated to be 23,000 lightyears across (this being four times Herschel's estimate). Hertzsprung's cepheid calibration put the SMC 100,000 lightyears away, suggesting that it and its larger companion were located *outside* the Milky Way.

In 1914 Shapley used the calibrated cepheid relationship to find the distances of the globular clusters. Because most of the cluster variables had periods shorter than the fastest in Hertzsprung's sample, Shapley sought longer period variables in some of the more readily resolved clusters, then used *these* to calibrate those with shorter periods, enabling him to measure the distance to a dozen nearby clusters. To measure the distance to those in which he could not identify variables, he assumed that their brightest members were intrinsically similar, and therefore could be used as 'standard candles'. Omega Centauri (the first of its kind to be recognised by Edmund

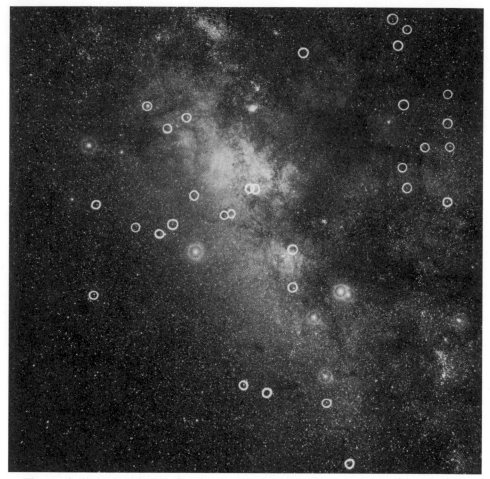

The profusion of globular clusters (circled) in Sagittarius. One-third of all known clusters are in this frame, which covers only about 2 per cent of the sky. The picture was taken with the 3-inch Ross–Tessar camera of the Harvard Observatory's Boyden Station in South Africa.

Halley in 1676) was found to be the nearest, at 20,000 lightyears. M13, the prominent cluster in Hercules, was at a distance of 36,000 lightyears. At 250,000 lightyears, NGC2419 in Lynx was the furthest. Despite these extreme distances, Shapley felt that they were associated with the Milky Way, and reasoned that if they were evenly distributed in *space*, and symmetrically distributed with respect to the Milky Way, then the Sun was nowhere near the system's centre, which was in the direction of Sagittarius. The band of the Milky Way is more prominent in that direction, but not so much as to hint that it encompasses the vast majority of the system's stars. Our view of the central region (and indeed, of the other side of the disk) is obscured. All the stars that we see in the sky are located on *one side* of the lens-shaped system. This explained why Herschel and Kapteyn, when counting stars, had underestimated the

The Horsehead nebula is part of a dense cloud of gas in front of the active star-forming nebula known as IC434 in the Orion complex alongside the three stars of Orion's belt. This picture was taken by T.A. Rector using the 0.9-metre telescope of the Kitt Peak National Observatory. Courtesy of NOAO/AURA/NSF.

size of the Milky Way and had inferred that the Sun was near its centre. In 1917 Shapley concluded that the system comprised a lens-shaped disk 300,000 lightyears across which was one-tenth as thick at its hub, around which the globulars clusters formed a spherical distribution whose diameter was comparable to that of the disk. In 1918 Shapley argued that the SMC was therefore *within* the confines of our system.

William Herschel dubbed dark patches in the Milky Way "holes in the heavens". When astronomers ventured into the southern hemisphere, they were struck by the dark 'Coalsack' which is set against the Milky Way. E.E. Barnard was a self-taught astronomer who, after many years as an amateur, studied at Vanderbilt University in Nashville, Tennessee, and in 1887 took a post at the Lick Observatory on Mount Hamilton, 3,000 feet above San José in California. In 1895 he was appointed professor of astronomy at the University of Chicago, where he had use of the 40-inch refractor of the Yerkes Observatory – the largest refractor in the world – and by 1919 he had catalogued 182 dark nebulae. The 'Horsehead' in Orion, which he inspected in 1913, was clearly a dark cloud viewed against a brighter nebula, rather than a region of space that was devoid of stars. In addition to dense clouds of luminous gas, it was evident that interstellar space was tenuously pervaded by cold gas and dust. In 1904 J.F. Hartmann identified absorption lines in starlight imprinted by such interstellar gas. Old red giant and supergiant stars shed carbon in the form of soot. A few grains per million cubic metres in interstellar space would significantly absorb and 'redden' starlight, giving the impression that the stars were further away than they actually were. In 1930 R.J. Trumpler at Lick Observatory published a study in which he measured the interstellar absorption of star clusters. The attenuation was greatest in the plane of the disk, and particularly in the direction of Sagittarius. Since most of the globular clusters were in this direction, Shapley had overestimated the distances of the more remote clusters and, in essence, had distended their gross distribution. Once this was realised, the Dutch astronomer J.H. Oort reduced the diameter of Shapley's disk to 100,000 lightyears. Its thickness in the solar

This 200-inch picture of NGC4565 illustrates what Shapley's model of the Milky Way would look like seen from far away and a perspective in the plane of the disk.

neighbourhood (about 27,000 lightyears from the centre) is 3,000 lightyears. With minor refinements, this scale has stood the test of time.

THE 'ISLAND UNIVERSES'

In 1882 J.L.E. Dreyer, one of Parsons's assistants, was appointed director of the Armagh Observatory in Ireland. He published the New General Catalogue (NGC) of 7,800 star clusters and nebulae in 1887, adding the Index Catalogue (IC) with another 5,386 in 1907. Isaac Roberts in England photographed M31 in Andromeda through his 20-inch reflector in 1886 using one of the new 'dry' plates and was astonished at the result. Up to that time, this fairly bright nebula had been described as a fuzzy oval, but the photograph showed it to be much larger, with its outer region possessing a prominent spiral form. It was possible that Parsons had not seen this structure because the field of view of his Leviathan had been so narrow. In 1899 Julius Scheiner in Germany secured a spectrum whose continuum implied that M31 comprised unresolved stars. Was it a small cluster lying within the confines of the Milky Way, or an independent system comparable to our own, and hence incredibly far away?

J.E. Keeler started a search for nebulae in 1898 by photographing selected areas of the sky using the 36-inch Crossley reflector at the Lick Observatory. By 1900 he had found ten times more than he had expected (most of them spirals) and concluded that in all there must be at least 100,000 – an estimate that was received with some scepticism. A successful English businessman named Percival Lowell set up home in Boston, USA, in 1638. Massachusetts lore says that the wealthy Cabots spoke only to the Lowells and that the Lowells spoke only to the Almighty. In 1855 this dynasty produced another Percival Lowell who graduated from Harvard where his brother was college president. After the customary tour of the world, Percival was assigned to run one of the family businesses, but he had developed such a passion for astronomy that, in 1893, he decided to establish his own observatory. With the assistance of W.H. Pickering of Harvard College Observatory, he bought a fine 24-inch refractor and selected as the site for his observatory the small sleepy railroad town of Flagstaff in Arizona which, by being on a plateau at an altitude of 7,200 feet, offered excellent seeing for much of the year. In 1901 V.M. Slipher, fresh from the University of Indiana, was hired and set to work investigating the spectra of nebulae. In 1913 he found that the spectrum of M45 was a continuum as a result of reflecting

When Isaac Roberts took this historic first photograph of M31 on 29 December 1888 he not only discovered the true extent of the nebula, but also that its outer region showed a spiral form.

the light of the Pleiades star cluster immersed within it. Critics of the idea that spirals were island universes argued that M31 was similar, but with denser gas masking the stars.

In 1885 C.E.A. Hartwig, a German astronomer at the Dorpat Observatory, saw a nova which, although it peaked at an unremarkable seventh magnitude, was of great interest because it appeared in the centre of M31. Very little was known concerning the nature of 'guest stars'. The nova that J.R. Hind saw in Ophiuchus in 1848 was the first to reach naked-eye visibility for 178 years, and it peaked at fifth magnitude. The one that was noted in 1866 in Coronae Borealis by John Birmingham in Ireland attained second magnitude. This patch of sky had been inspected four hours earlier by J.F.J. Schmidt at the Athens Observatory, so whatever the object might be, it had brightened by at least four magnitudes in that interval. It was the first to be studied spectroscopically. A few days after its peak, William Huggins discovered that as the outburst faded the spectrum changed to that of a cloud of gas that was heated from within, with bright emission lines superimposed upon a continuum.

The 'Firework' nebula is the shell of material ejected by the nova in Perseus in 1901. This image was taken in 1994 with the WIYN 3.5-metre telescope on Kitt Peak. The WIYN Consortium comprises the University of Wisconsin, Indiana University, Yale University, and the National Optical Astronomy Observatories (NOAO). Courtesy of N.A. Sharp.

The nova that the Scottish clergyman/astronomer T.D. Anderson discovered in Perseus in 1901 was the first to be observed while still on the rise. It eventually attained zeroth magnitude. A picture taken by M. Wolf six months later using a 16-inch astrograph showed an arc of nebulosity nearby. The radial velocity of the emission lines indicated that the cloud of gas was being ejected at about 1,600 kilometres per second. More detailed pictures taken by the Yerkes and Lick telescopes indicated a progressive drift of the arc away from the star. The fact that the nova had been so prominent had prompted speculation that it was not too far distant. This impression was reinforced by correlating the radial velocity with the rate of increase of the angular diameter of the nebulosity. A distance of several hundred lightyears was estimated. In 1911 F.W. Very in America calculated that it would have to have been set at a distance of several thousand lightyears to have appeared as faint as the 1885 nova, placing that object well inside the Milky Way. If the nebulosity of M31 was further away, then the nova was simply on the same line of sight. In 1916 W.H. Steavenson and E.E. Barnard independently noticed another nebulosity close to the 1901 nova which, by 1920, was shown to be expanding at the indicated radial velocity. J.C. Kapteyn explained the earlier effects as the flash propagating through a pre-existing dark cloud and illuminating dense strands of gas and/or dust. Once it was understood that *this* angular rate corresponded to the speed of light, the distance estimate could be revised upward to about 1,500 lightyears. As the distance to the 1901 nova had been grossly underestimated, so too had Very's derivation of the distance to that of 1885. This reopened the debate on whether the alignment had been a matter of chance.

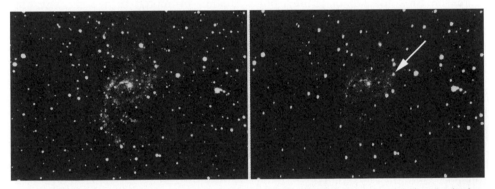

In 1917, while comparing two plates of NGC6946 (a spiral in Cepheus) that he had taken using the 60-inch telescope at the Mount Wilson Observatory, G.W. Ritchey discovered what appeared to be a nova out in its peripheral region. This was the first nova to be spotted photographically.

To those who thought the spirals to be Laplacian nebulae, the fact that the 1885 event had been centrally located in M31 was interpreted as the newly formed star in the heart of the collapsing cloud 'switching on'. Its subsequent fading was explained as a swirl of dust masking the star. In 1917 G.W. Ritchey at the Mount Wilson Observatory was comparing two plates by the 60-inch telescope of NGC6946, a spiral in Cepheus, when he noticed that a star in its *outer fringe* was present on one plate and absent on the other. In this peripheral location it could not be a newly formed star in a Laplacian nebula. In a search of the archive, Ritchey found two such apparitions in M31 on plates that he had taken in 1909. After joining the Lick Observatory in 1902, H.D. Curtis began to photograph spirals. Upon hearing of Ritchey's discovery, he searched the Lick archive and found a nova in NGC4227 in Canes Venatici and two in M100 in Coma Berenices. Other observatories also examined their plates. Within two months of Ritchey's announcement, 11 novae had been discovered, four of them in M31. At their peak these were much fainter than that of 1885. Confident that the four in M31 were *in* the spiral, Curtis realised that the 1885 event must have been either a chance alignment or a much more powerful outburst. He calculated that for the 1901 nova to have appeared as faint as the four in M31 it would need to have been removed to a distance of 500,000 lightyears, which would make M31 a vast system of millions of stars. Shapley estimated that the greatest distance at which a nova could be detected was 8,000 lightyears, and noted that if the 1885 nova had been in M31 – and as remote as Curtis had said – then that outburst would have to have outshone the entire system, which Shapley insisted was absurd.

After leaving school without qualifications in 1906, M.L. Humason was hired as a mule train-driver to take stores from Los Angeles up Mount Wilson to build the observatory. In 1911 he married the daughter of the observatory's chief engineer. As late as 1914, the only access was by way of a 15-kilometre-long path that wound its way slowly up to the 6,000-foot summit, and *everything* went up by mule. In 1917 Humason took the job of janitor, but was soon promoted to a night assistant. Working for Shapley, he taught himself sufficient astronomy to pursue research of his own, and in 1919 became a junior member of the observing staff. When he spotted what seemed to be a variable star in M31, he took the plates to Shapley, who insisted that the transient patch of light was a knot of matter within the swirling Laplacian nebula.

In 1914 E.P. Hubble took a job as an assistant at the Yerkes Observatory while he studied for his doctorate at the University of Chicago. After graduating in 1917 he served in the infantry during the Great War in Europe and on his return in 1919 he was hired by the Mount Wilson Observatory. Hubble used the newly commissioned 100-inch telescope to document the two spectacular spirals: M31 (which is viewed obliquely) and M33 in Triangulum (which is 'face on'). When he remarked that the glow in their outer regions was reminiscent of low-resolution pictures of the densest star fields in the Milky Way, Shapley insisted that the glow was nebulosity. On 26 April 1920 the Smithsonian Institution in Washington hosted a debate on *The Scale of the Universe* in which Shapley and Curtis argued about the size of the Milky Way system and whether the spirals were independent systems comparable to our own. Although each man considered himself to have won the argument, in the absence of convincing evidence the result was actually a draw. After the death of E.C. Pickering in 1919, Shapley had been invited to succeed him as director of the Harvard College Observatory. By the time of the Great Debate, his initial reticence had eroded, so a few months later he moved to America's first observatory, which had become an astronomical auditing house that compiled vast catalogues. With the most powerful telescopes in the world, Mount Wilson was the base camp for celestial *explorers*. In his memoirs, Shapley described his years at Harvard as "anticlimatic".

M33 in Triangulum is a large spiral with a negligible nuclear 'bulge'. This picture was taken with the 4-metre Mayall Telescope at the Kitt Peak National Observatory. Courtesy of NOAO/AURA/NSF.

The 100-inch reflector of the Mount Wilson Observatory constructed by G.E. Hale and commissioned in 1917 was funded by J.D. Hooker.

HUBBLE'S BREAKTHOUGH

As a preliminary to studying spirals, Hubble examined emission and reflection nebulae in order to verify what had already been inferred, and in 1922 he set out to determine whether spirals were part of, or beyond, the Milky Way system. After starting his career at the Royal Greenwich Observatory, C.D. Perrine had moved to the Lick Observatory in 1893. On observing NGC6822 in Sagittarius, he concluded that it was a ragged collection of stars, similar to the Magellanic Clouds. As the 60-inch reflector could resolve stars in NGC6822, Hubble launched a search for cepheids. Although by early 1923 he had discovered a dozen variables, the fact that he could see them only when near their maxima made it difficult to identify their types.

In the fall of 1923, upon acquiring an improved emulsion, Hubble took a plate of M31 through the 100-inch and was able, for the first time, to resolve stars in its periphery. Remarkably, this first plate caught two novae and a third object which, upon further monitoring, proved to be a variable. By the turn of the year, he had the unambiguous light curve of a cepheid with a period of about a month. With this, he

The first nebula in which Hubble noted variable stars was the irregular NGC6822 in Sagittarius.

calculated that M31 was 8.5 times more distant than the SMC. Using Hertzsprung's estimate to the SMC of 100,000 lightyears, this put M31 at 850,000 lightyears, at which distance its angular diameter meant that it must rival the Milky Way system. In February 1924 Hubble penned Shapley a brief note in which he expressed his confidence that Shapley would "be interested to hear that I have found a cepheid in the Andromeda nebula". C.H Payne, a specialist in variable stars, was with Shapley when he received the note. He wryly told her that Hubble had just destroyed his 'universe'. In his reply, Shapley described Hubble's note as "entertaining literature". (During their brief time together on Mount Wilson, Hubble and Shapley had found each other's demeanor irritating. When an editor asked him to referee a paper that Shapley had sent to a journal, Hubble neatly inscribed 'Of no consequence' across it. The paper was accepted, however, and for some reason the typesetter inserted Hubble's dismissive remark just below the author's byline!) In a paper for a meeting of the American Association for the Advancement of Science in Washington, DC, held on 1 January 1925, Hubble reported having identified a dozen cepheids in M31. When he returned to NGC6822, he used the larger telescope's greater light grasp to follow its variables in detail. After identifying a number of cepheids, he calculated that it was 700,000 lightyears away. A tally of 35 cepheids implied that M33 was

This magnificent photograph of M31 in Andromeda and its small companions M32 (NGC221; lower centre) and M110 (NGC205; upper right) was taken in 1979 by Vanessa Harvey using the Burrell Schmidt Telescope on Kitt Peak operated by the Case Western Reserve University's Warner and Swasey Observatory. The box delineates the area covered by the next image. Courtesy of Bill Schoening, NOAO/AURA/NSF.

Two of the cepheids spotted by E.P. Hubble in the periphery of the M31 spiral.

850,000 lightyears away. Hubble published a seminal paper on M33 in 1926, and one on M31 in 1928. A few years with the new telescope had proved that the spirals really were 'island universes'. In recognition, the Greek word for 'milky' was coined, initially in the form of 'extragalactic nebulae', but later, in acknowledgement that they are not nebulae at all, this was contracted to 'galaxies'.

In 1910 Lowell, who believed the spirals to be Laplacian nebulae in the process of forming planetary systems, had directed Slipher to attempt to measure their radial velocities in the hope of determining their rate of rotation. Lengthy exposures were required simply to photograph spirals, and once their feeble light was dispersed with sufficient resolution to measure a radial velocity, it would take much longer to record a spectrum. It took two years of trials to devise a viable method, and even the most prominent spiral required 'integrating' light over several nights. M31 was found to be approaching at 300 kilometres per second. At the time, this was the largest radial velocity of any celestial object measured. Why should a spiral be moving so rapidly? By 1914, when Slipher sent his results to the American Astronomical Society, all but two of the 15 spirals he had measured were *receding* at even higher speeds. Why were they moving in such a systematic manner? The mean recessional velocity of 600 kilometres per second supported the case that they were beyond the Milky Way system. By 1923 Slipher had 41 radial velocities, and other observatories had added four more. The greatest recessional velocity was an astounding 1,800 kilometres per second. On Lowell's death in 1926, Slipher was made director of the renamed Lowell Observatory and he turned to other projects. In any case, he had pushed the 24-inch refractor to its light-gathering limit. In 1924, C.W. Wirtz in Germany inspected the spirals on Slipher's list and noted that if they were assumed to be of similar size, then those that were further away would appear smaller. He also suspected that there was a relationship between distance and recessional velocity, but this required the distance of each spiral to be determined and he did not have the facilities to test his hypothesis. Even with the 100-inch, cepheids could be seen only in the nearest of galaxies, and Hubble had secured distances to only six galaxies using them. In order to probe further, he assumed that the brightest stars in any galaxy were of similar luminosity and, like Shapley before him with the globular clusters, employed these as standard candles. When he was no longer able to resolve stars, he assumed that galaxies of similar form were similar in luminosity. In the absence of evidence to the contrary, these were fair assumptions. Hubble reasoned that he could see the dwarf galaxies in our neighbourhood precisely *because* they are nearby. If M31 were to be viewed from a much greater distance, its dwarf companions would fade into obscurity. On noticing a 10-fold difference in brightness between the most and least prominent members of a cluster of galaxies in the direction of Virgo, Hubble inferred that he was seeing only the *large* galaxies. On the assumption that these were equally luminous and were evenly distributed within a spherical cluster, he calculated how far the galaxies for which he was able to derive distances would have to be in order for them to seem as faint as the mid-point in this brightness range, and thereby calculated the distance of 6.6 million lightyears.*

Ever since Herschel had realised that proper motions meant that the nearby stars shared the Sun's motion, it was presumed that this was because the system of stars was in a state of rotation. In 1926 J.H. Oort estimated that the Sun was moving at 250 kilometres per second in a 200-million-year orbit. Once this systematic bias was eliminated from Slipher's radial velocities, the reason for most of M31's speed of approach was evident: at this point in its orbit, the Sun is moving in that general direction. The realisation that M31 is actually approaching at only 50 kilometres per second made the recessional velocities of all the others seem even more significant. By early 1928 Hubble had measured the distances of 24 of the spirals on Slipher's list. When he plotted their distances and velocities he found, as Wirtz had suspected, that there was a linear relationship. On average, the velocities increased by about 160 kilometres per second per million lightyears of additional distance. Lest this rate sound wild, in more familiar terms it corresponds to the separation between two objects 1,000 kilometres apart increasing by a fraction of a millimetre per year. Nevertheless, the clear implication was that the Universe was expanding. In fact, the *geometrical* relationship meant that the spirals were racing away from one another *at a constant rate*. Only two were approaching, and the fact that M31 was one of the nearest suggested that it was gravitationally bound to our own. In 1929, in a paper entitled 'A Relation between Distance and Radial Velocity among Extragalactic Nebulae', Hubble published his astounding conclusion that in addition to being considerably larger than had been thought, the Universe was also in a state of expansion.

---

*   Intergalactic distances are quoted in the context of *contemporary* estimates of the rate at which the Universe is expanding.

8

Cosmology

AMAZING IDEAS

Upon graduating from Cambridge University in 1907, A.S. Eddington joined the Royal Greenwich Observatory. He became professor of astronomy and experimental philosophy at Cambridge in 1912. After studying astronomy under J.C. Kapteyn at the University of Gröningen in the Netherlands, Willem de Sitter spent two years at the Cape Observatory before returning to the Netherlands to gain his doctorate in 1901 at the University of Leiden, where he was appointed both professor of theoretical astronomy and director of the Leiden Observatory in 1908. Eddington and de Sitter had corresponded regularly on issues arising from Albert Einstein's Special Theory of Relativity, and although the Great War was raging when Einstein's paper 'The Foundations of the General Theory of Relativity' was published in *Annalen der Physik* in Germany in 1916, the Netherlands was neutral in the conflict and de Sitter was able to mail a copy to Eddington. When Einstein published his paper 'Cosmological Considerations on the General Theory of Relativity' in 1917, this was also forwarded to England. Einstein's astronomy was based on Kapteyn's paper of 1906, reporting that the Sun was situated near the centre of a disk of stars. Shapley was studying the distribution of globular clusters, but he had not yet published. As yet, there was no conclusive evidence that the spirals were 'island universes'. Slipher was finding that the spirals had large radial velocities, with a surprising trend, but Einstein was not aware of this. The Milky Way system was believed to be in a state of rotation, but the radial velocities listed in star catalogues did not show a systematic expansion or contraction. Although the Universe was believed to be static, the equations led naturally to solutions in which it was evolving – either expanding or contracting. Einstein therefore reasoned that there must be a force acting to counter the tendency of the Milky Way system to collapse under the mutual gravitational attraction of its stars, and he introduced the 'cosmological constant' into his equations to represent a repulsive force that applied at long range. In a way, Einstein was unlucky: his theorising had run ahead of the observational evidence, and like all mathematical physicists he tried to make his theory 'fit'. His first exposure to the 'island universe' theory may have been the debate between Shapley and Curtis in 1920, which he attended, but its outcome was inconclusive. If he had introduced his theory a decade later, then it would have dovetailed nicely with Hubble's findings, and their papers could well have appeared as a *tour de force* in the same journal.

Einstein had expected there to be just one solution to his equations, and having found it he studied its implications, but in 1917 de Sitter reported another solution in a paper entitled 'On the Curvature of Space'. Later in that year, in the paper 'On Einstein's Theory of Gravitation and its Astronomical Consequences', he argued that the stars are so far apart that the overall matter density is insignificant, and for his study he presumed it to be zero. He retained the cosmological constant because he sought a static solution. He initially thought that he *had* produced a static solution, but when Eddington introduced mathematical 'test particles' these diverged because the *space* that the equations described was in a state of expansion. In response, de Sitter quipped that it was a static solution in the sense that there was no matter to *reveal* its expansion! Einstein dismissed de Sitter's solution as lacking physical significance: the Universe was not 'empty', and it was the *presence* of matter that curved space. Eddington was intrigued that Einstein could identify a solution in which there was matter but no expansion while de Sitter found an empty one that was nevertheless in a state of expansion.

In 1910 A.A. Friedmann had joined the mathematics faculty at the University of St Petersburg. After the Great War and the Revolution he took a teaching post at the University in Perm. In 1920 he returned to St Petersburg, now Leningrad. Although a meteorologist, he was interested in Einstein's work. In 1922 he ignored the focus upon static solutions and studied a family of expanding solutions, one being 'open', one 'flat', and the third 'closed', with the matter density distinguishing them, and he expressed them with and without a cosmological constant. He calculated the critical density for a 'flat' Universe as 10^{-28} g/cm^3, which is equivalent to an average of only a few hydrogen atoms per cubic metre. Upon being told of this study, Einstein was amazed that his theory could generate so many solutions. It had been presumed that the Universe was eternal and unchanging, but expansion appeared to indicate a point of 'origin' at 'time-zero'. Unfortunately, in 1925 Friedmann caught a severe chill while setting an altitude record of 23,000 feet in a balloon, succumbed to typhus and died. Eddington missed his papers in *Zeitschrift für Physik*. During the Great War, Georges Lemaître served as an artillery officer in the Belgian army. After the war he attended the University of Louvain and was ordained as a priest in 1923. While working for his doctorate he spent a year with Eddington in Cambridge and a year in Boston, attending Harvard and the Massachusetts Institute of Technology. He was appointed professor of astronomy upon returning to Louvain in 1927. In America he had heard of Slipher's radial velocity measurements, and when he independently devised Friedmann's solutions he readily accepted that the Universe was expanding. However, he avoided the issue of an 'origin' by retaining the cosmological constant and arguing that, after spending a long time in the static state envisaged by Einstein, an instability had initiated a runaway expansion. Unfortunately, Lemaître published in the *Annals of the Brussels Scientific Society*, and once again Eddington missed the paper.

THE MOST AMBITIOUS PROJECT

At an International Astronomical Union symposium on galaxies in the summer of 1928, de Sitter met Hubble, and they put two and two together. Later in the year, W.S. Adams, the director of the Mount Wilson Observatory since Hale's retirement in 1922, gave priority on the 100-inch to Hubble to extend the 'depth' of his survey. "This is the most ambitious project astronomy has ever known. Maybe you'll reach out to the very edge of the Universe!" Hubble teamed up with Humason, and in the division of labour Hubble estimated the distances to galaxies employing a 'ladder' of 'indicators' and Humason took spectra to measure their velocities. However, it was becoming increasingly difficult to probe deeper – Humason had to integrate for *ten* nights to secure the spectrum of a galaxy in a cluster in Ursa Major. Nevertheless, in a joint paper entitled 'The Velocity–Distance Relation among Extragalactic Nebulae', published in 1931, they spectacularly extended the linear relationship to a speed of 20,000 kilometres per second, and out to an estimated distance of about 100 million lightyears.

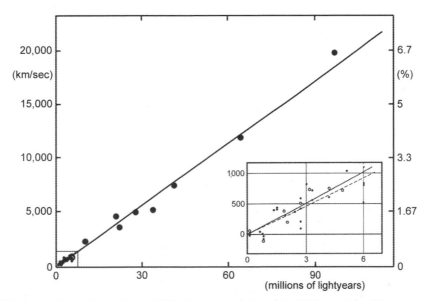

The inset shows the initial redshift–distance relationship (with considerable 'scatter' in the distance estimates) published by E.P. Hubble in 1929. The main plot shows the greatly extended relationship he derived with M.L. Humason and published in 1931. The redshift is also expressed on the right in terms of the percentage shift in spectral lines.

THEORY CATCHES UP

In January 1930 Eddington told a meeting of the Royal Astronomical Society that in light of Hubble's discovery of the redshift–distance relationship, static solutions to Einstein's equations were inappropriate. When G.C. McVittie, one of his research students, discovered Lemaître's paper, Eddington arranged for this to be reprinted in the *Monthly Notices of the Royal Astronomical Society* and followed with a paper of his own in which he discussed the implications. By this point, however, Einstein had renounced his cosmological constant, which he derided as his "greatest blunder", and Lemaître, following suit, had come to the conclusion that all the mass of the Universe must once have been concentrated into a 'cosmic egg'. The process of radioactive decay was barely understood, but he envisaged the decay of a superdense 'primeval atom' as being analogous. Lemaître published this bold idea in 1931 in a paper entitled 'The Beginning of the World from the Point of View of Quantum Theory' in the journal *Nature* to ensure that it reached the widest possible audience. In 1932 Einstein and de Sitter wrote jointly 'On the Relation between the Expansion and the Mean Density of the Universe', in which they embraced the work of Friedmann and Lemaître and identified a 'flat' solution that was consistent with expansion and in which the cosmological constant was zero (this 'Einstein–de Sitter Universe' was actually a special case of the simplest of Friedmann's solutions). The theory accounted for the illusion of the redshift–distance relationship that we inhabit a privileged position at the centre of the Universe – the galaxies were not receding *through* space at the high speeds indicated by their radial velocities, it was *space* that was expanding and the galaxies were being carried along, with a geometrical effect making those further away *appear* to recede more rapidly. In December 1932 Lemaître visited Caltech and gave a seminar on the 'cosmic egg'. Einstein, who was on sabbatical leave in America, stood up afterwards and announced, "This is the most beautiful and satisfactory explanation of creation to which I have ever listened." When Hitler rose to power a few months later, Einstein decided to remain in America, and took a professorship at the newly established Institute for Advanced Study at Princeton University in New Jersey.

THE BIG EYE

By 1936 Hubble and Humason had measured some 150 velocities, ranging up to 40,000 kilometres per second in a survey that encompassed a *vast* volume of space. When J.E. Keeler had conducted a photographic survey at the turn of the century he ventured that there might be as many as 100,000 galaxies. Hubble revised this figure to 100 million. As recently as 1918, the Milky Way had been believed to comprise the entire Universe. The motion of the Sun indicated that the gravitating mass of the galaxy was 100 billion solar masses, and as most stars are significantly less massive than the Sun this indicated that it contained approximately *200 billion stars*. Despite this, our system had been shown to be just one of a myriad of galaxies.

Walter Baade received his doctorate in 1919 from the University of Göttingen and spent the 1920s at the Bergedorf Observatory in Hamburg. In 1931 he moved to the

Mount Wilson Observatory. As Hubble and Humason probed deeper into space, Baade suspected that there was a flaw in the distance scale. Why should our galaxy be larger than the others? The globular clusters in M31 were spherically distributed around its nucleus, just as Shapley had discovered for those associated with our own system, but why were they only half as luminous as ours? And why were there no short-period variables? The age of the Universe could be calculated by running the expansion in reverse, but the value of the rate of expansion inferred by Hubble meant that the Universe was 2 billion years old, which conflicted with Ernest Rutherford's radioactive dating of 3.4-billion-year-old rocks. As he mused these mysteries, Baade made a start on his own research. As a student he had been impressed by the fact that because the Hertzsprung–Russell diagram identified well-defined types of star it "opened up an entirely new way of studying galactic structure". As a preliminary to studying our own galaxy, he decided to determine the distribution of different types of stars within a spiral. The obvious candidate was M31. Hubble had tried to resolve stars in its nucleus, but his plates had shown only a glowing mass, and Baade found likewise. He tried again when Los Angeles was 'blacked out' early in the Second World War, still without success. Like Hubble, he had used a blue-sensitive emulsion because it was 'faster'. However, the air molecules scatter most strongly at the blue end of the spectrum – which is why diffused sunlight is blue – and there was a faint background glow in the night sky even during the blackout. In an effort to reduce this interference, Baade switched to a red-sensitive emulsion, but because this was much 'slower' he had to use much longer exposures. His long nights at the telescope were rewarded with resolved stars in M31's nucleus. Hubble had been drafted as an 'army scientist'. Classified as an enemy alien, Baade was restricted to the Observatory for the duration, but he was content because he had the 100-inch more or less to himself and was finally able to push it to its theoretical limit. He found that whereas the peripheral regions of M31 were predominantly 'hot blue' stars, the brightest stars in the nucleus were reddish. Furthermore, while the spiral arms were laden with gas and dust, the nucleus was relatively clear, and he therefore inferred that there were two 'populations'. In terms of the evolutionary sequence represented by the Hertzsprung–Russell diagram, the spherical nucleus and the surrounding globular clusters were composed of old stars, which were in the red giant or supergiant stages of their evolution. The blue stars in the spiral arms of the flattened disk were younger. In effect, because hot blue supergiants were short-lived, the blue stars that must have formed in the inner regions had long since expired. The fact that the nucleus was clear of dust meant that the raw material from which stars formed had been used up, and star formation had ceased. In the dusty spiral arms, in contrast, the process was ongoing. The difference in ages of the populations was evident from the fact that the young hot stars showed more 'metals' in their spectra, indicating that they formed from clouds that had been enriched by earlier generations of stars.* When Baade resolved several elliptical

---

* For some reason, Baade named the populations in order of their discovery rather than in order of their formation, so the old red stars in the nucleus became Population II and the young blue stars in the spiral arms became Population I.

Striving to probe deeper into the Universe, in 1928 G.E. Hale secured funding from the Rockefeller Foundation for a 200-inch reflector on Mount Palomar. Upon being commissioned in 1949, it was dedicated to Hale, who died in 1938. However, local astronomers informally referred to the Hale Telescope as the 'Big Eye'.

galaxies in 1944 he found them to be old red stars, implying that such galaxies were analogous to the nuclear regions of spirals.

After extending the redshift–distance relationship as far as the 100-inch would allow, Hubble had to bide his time during the protracted development of the Hale Telescope. In 1948 he took a test picture through the new telescope even before the engineers had finished refining its 200-inch mirror, and was delighted to verify that a 10-minute exposure could capture faint galaxies which the 100-inch was only barely able to record by integrating all night. In 1949 Baade's first plate of M31 confirmed his suspicion that the distance scale was flawed. At a distance of just under 1 million lightyears he ought to have been able to resolve short-period variables but they were absent at the expected magnitude, so either such stars were less luminous in M31 or it was much further away than believed. By 1952 Baade was able to announce at the International Astronomical Union meeting in Rome that there were two types of cepheid, with the 'classical' ones such as delta Cephei being 1.5 magnitudes more luminous than their metal-poor cousins, the W Virginis variables. Applying this correction increased the distance of M31 to 2.2 million lightyears, and eliminated the issue of the brightness of its globular clusters. Hubble had measured the slope of the redshift–distance relationship as 160 kilometres per second per million lightyears. In halving this, Baade had doubled the radius of the known Universe and increased its age sufficiently to accommodate the oldest known rocks. There was still a problem, however, because stellar theory suggested that the old red stars in globular clusters were even older. In his survey of M31, Baade found that he could trace the spiral arms by hot blue stars, and he suggested that a similar study should reveal the spiral structure of our own system. In 1951 W.W. Morgan at the Yerkes Observatory used emission from ionised hydrogen in the vicinity of newly formed hot blue stars to show that the Sun is situated near the inner edge of the Orion Arm, and he was able to trace sections of the outer Perseus Arm and the inner Sagittarius Arm. As most of the stars visible to the naked eye are within these local structures, we see only a very small part of the galaxy – as Herschel and Kapteyn had unwittingly found.

EXPLODING STARS

After a decade of positional tracking, F.W. Bessel had realised in 1844 that both Sirius and Procyon 'wobbled' as they pursued their proper motions, from which he inferred that they had unseen companions. What could be sufficiently massive as to perturb a star and yet remain invisible? A.G. Clark inspected Sirius in 1862 while testing the 18-inch refractor (the largest lens in the world at that time) that was destined for the University of Chicago's Dearborn Observatory, and was astonished to *see* the companion as an 11th-magnitude star almost lost in its primary's glare. In 1896 J.M. Schaberle saw Procyon's companion using the Lick Observatory's 36-inch refractor.

The white dwarf Sirius B is an 11th-magnitude star almost lost in its primary's glare.

The companions were presumed to be very old and slowly cooling stars. W.S. Adams at the Mount Wilson Observatory isolated Sirius B's spectrum in 1914 and eventually identified highly ionised lines implying a temperature of 10,000 K. To be both hot *and* faint, the star had to be physically small – in fact, no more than 20,000 kilometres in diameter. By some process, the mass of the star had been forced into a volume comparable to that of the Earth. In 1926 R.H. Fowler in Cambridge investigated this extremely dense state. A star is an evolving balance between gravitational collapse and the pressure of the radiation created by the nuclear processes in its core. Once the source of energy is exhausted, gravity finally takes its toll, squeezing the star to a density of 10^9 g/cm^3, forcing the atoms into physical contact and their electrons into their lowest states. Further collapse is prevented by Pauli's Exclusion Principle, which manifests itself as the pressure of this 'degenerate' electron gas. As the core of a red giant contracts to form such a 'white dwarf', the energy released ejects its envelope, creating a planetary nebulae (which solved another mystery). While sailing to England in 1930 to work for his doctorate under Fowler's supervision, Subrahmanyan Chandrasekhar, a physics graduate from Madras University in India, calculated that there was an upper limit of 1.44 solar masses for a white dwarf. It therefore followed that as a more massive star neared the end of its life, the pressure of the degenerate electrons in its core would not be able to resist total gravitational collapse. Eddington rejected 'Chandrasekhar's limit' (as it became known) because it implied that a star could cut itself off from the surrounding space and cease to exist, which he thought was ludicrous. After securing his doctorate in 1933 Chandrasekhar went to America. In 1936 he accepted a post at the University of Chicago, where he promptly established himself as the world's leading theorist on stellar structure.

When astronomers began to take a serious interest in novae in the late nineteenth century, no one knew what they were. In 1879 A.W. Bickerton proposed that they were the result of close encounters (or even collisions) between stars. This led to the suggestion by T.C. Chamberlain and F.R. Moulton in 1901 that the Earth and the other planets condensed from a streamer of gas that was torn out of the Sun during such an encounter – and the appearance of nebulosities around novae seemed to support this. On the other hand, W.H.S. Monck and Hugo von Seeliger had independently suggested in 1892 that the sudden increase in brightness was due to a dark body (not a star) passing through a cloud of interstellar gas. It was suggested that the putative frictional heating explained both the continuum and the emission lines in the spectrum. A variety of theories were proposed but, even as late as the mid-twentieth century, most focused upon notional 'instabilities' in stars. The favoured model is a close binary system in which hydrogen from a main sequence star is being accreted by a white dwarf, with it accumulating on the surface until it explosively ignites.

As Hubble explored M31's outer regions using the 100-inch, he spotted dozens of novae whose faintness confirmed that if the 1885 outburst had been in this galaxy then it must have been considerably more powerful. After obtaining his doctorate in Switzerland in 1922, Fritz Zwicky relocated to America in 1925 and joined Caltech. In 1934 he began searching for such outbursts in galaxies. In 1930 B.V. Schmidt at the Hamburg Observatory had designed a telescope capable of photographing a wide field. Baade had taken this plan with him when he moved to the Mount Wilson Observatory and Zwicky, upon being shown it, secured funding for an 18-inch Schmidt on Mount Palomar, where development work for the 200-inch telescope was already underway. With a field of view equivalent to an area the size of the

'Plough' of Ursa Major, it could record a large number of galaxies on a single plate. Within two years of it being commissioned in 1936, Zwicky had discovered a dozen 'supernovae' (as he named them). R.L.B. Minkowski, who had also worked with Baade in Hamburg, fled Germany in 1933 and then joined the Mount Wilson Observatory in 1935. Each time Zwicky discovered a supernova during one of his 'patrols', Baade monitored its light curve and either Humason or Minkowski secured its spectrum. At their peak, these outbursts *outshone* their host galaxies. As the list grew, Minkowski classified supernovae by their spectra at maxima and the manner in which they faded. He designated as Type I those that did not show strong lines of hydrogen (i.e. the majority, about 75 per cent) and as Type II those that did. Later studies introduced subdivisions.

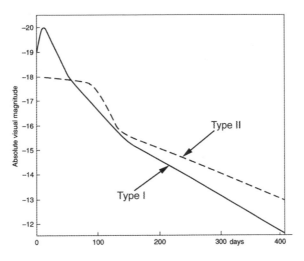

R.L.B. Minkowski classified supernovae by their spectra at maxima and the manner in which they faded. Type I do not show strong lines of hydrogen, whereas Type II do. Later studies introduced subdivisions. Type Ia supernovae have a strong silicon line. Other Type I supernovae are Ib if they show strong helium lines and Ic if they do not. The absence of hydrogen lines implied that the progenitors were evolved stars that had ejected their hydrogen envelopes. Another clue to their nature could be inferred from their locations. Type Ia were found in all kinds of galaxy, including ellipticals, and so were unlikely to mark the death throes of massive young stars. On the other hand, Types Ib, Ic and II were seen *only* in spiral arms near stellar nurseries, and so probably were the detonation of massive stars whose collapsing cores (depending upon their mass) formed either neutron stars or black boles. However, Type Ia are believed to be close binary systems in which a white dwarf accretes matter from its main sequence companion until its overall mass approaches the Chandrasekhar limit. At this point it undergoes a runaway thermonuclear reaction which ejects its outer layer at 10,000 kilometres per second to form a nebulosity that radiates strongly in the visible spectrum for a time as its radioactive elements decay. Detailed numerical models accurately explained the kinks in the light curve. The gas initially glows from the decay of nickel-56 to cobalt-56 with a half-life of 6 days and then, after a rapid decline, the slope follows the decay of cobalt-56 to iron with a 78-day half-life.

The Chinese chronicle *History of the Sung Dynasty* recorded the appearance of a 'guest star' in Taurus on 4 July 1054 that slowly faded from view over the next 653 days. A similar event had been observed in 1006, and another was seen in 1181, but there was then a long period with no such sightings. On 11 November 1572 a 'new star' appeared in Cassiopeia. It rivalled Venus for several weeks before progressively fading and disappearing in early 1574. When nobleman Tycho Brahe reported his observations in a short book *De Nova Stella* in 1573, King Frederick II of Denmark was so impressed that he gave Brahe the small island of Hven in the channel between Copenhagen and Helsingfors to establish an 'observatory' for a systematic study of the motions of the planets. His assistant, Johannes Kepler, studied another such star that flared in Serpens in October 1604. Its appearance stimulated Galileo Galilei's interest in astronomy. In retrospect, these were all supernovae. 'Kepler's Star' and 'Tycho's Star' were Type I; that of 1054 was Type II. Whatever novae might be, it was evident that supernovae were orders of magnitude more powerful, and the only plausible process was the *catastrophic explosion* of a star.

THE PRIMORDIAL FIREBALL

George Gamow's interest in cosmology was stimulated by attending lectures by Friedmann at the University of Leningrad in 1924 while working for his doctorate, which he gained in 1928. He then embarked on a tour which took him to the University of Göttingen (where he used quantum theory to explain how an alpha particle could 'tunnel' out of an unstable atomic nucleus), the Institute of Theoretical Physics in Copenhagen and Cambridge University in England. On his return to Leningrad in 1931 he was so depressed by the deterioration in the political situation that while on another tour in 1933 he decided not to return and made his way to America where he took up an appointment at George Washington University in Washington, DC.

Gamow endorsed Lemaître's idea that the Universe emerged explosively from a superdense 'primeval atom'. C.H. Payne, while studying at Cambridge University, was encouraged by Eddington to pursue astrophysics, and after graduating in 1923 she went to the Harvard College Observatory where Shapley set her the task of analysing stellar spectra. Her doctoral thesis in 1925, 'On stellar atmospheres', was described as being "brilliant". In addition to concluding that the relative abundances of the elements in stars were similar to those in the Sun, she introduced the term 'cosmic abundances'. Lemaître's fission theory could not explain the fact that hydrogen was predominant. As an alternative, Gamow suggested that Lemaître's 'atom' produced only neutrons, and because free neutrons are unstable these would have progressively decayed into equal numbers of protons and electrons. Gamow argued that the nuclei of the heavier elements were created by the fusion of protons with the late-decaying neutrons. He announced his theory in the summer of 1935 at a seminar at Ohio State University, and followed it with a paper in the *Ohio Journal of Science* entitled 'Nuclear Transformations and the Origin of the Chemical Elements'. While his predecessors had treated cosmology as a study in geometry, Gamow studied it in terms of high-energy physics. While working as a consultant to the Applied Physics Laboratory of Johns Hopkins University in Baltimore during the Second World War, Gamow continued to develop this cosmological theory, and in 1946 wrote a paper for *Physical Review* entitled 'The Expanding Universe and the Origin of the Elements'. When R.A. Alpher abandoned his doctoral work on cosmic rays upon finding that he was a year or so behind a rival at another institute, Gamow agreed to supervise him in a study of fusion processes, exploiting recent experimental results on the process of neutron capture by D.J. Hughes at the Argonne National Laboratory in Chicago. Although Alpher's calculations confirmed that helium would be produced, he found that the process stalled at this point. Even as Alpher was writing his thesis, his conclusions were submitted for publication. As Alpher had done most of the work, he was the leading author. Before submitting the paper to *Physical Review*, Gamow, a renowned prankster, mischievously added the name of H.A. Bethe at Cornell as second author, parenthesised 'in absentia' so that the list would read like the first three letters of the Greek alphabet. Unbeknown to Gamow, the draft was sent to Bethe for comment, who endorsed it by eliding the 'in absentia', and making himself an author of the seminal paper, 'The Origin of the Chemical Elements'. As if to reinforce the jape, that particular issue of the journal was issued on 1 April 1948!

Though the idea of 'primordial nucleosynthesis' was flawed, it could explain the preponderance of hydrogen and helium and so, as Gamow delightfully observed, it successfully accounted for *99 per cent* of the matter in the Universe. Furthermore, the study had revealed that the neutron capture process was potentially so efficient that it could easily have turned *all* the hydrogen into helium. The helium abundance therefore implied that there must have been a radiation field that acted to break up the helium as it was formed, with the outcome being the result of an equilibrium between all these processes. Alpher teamed up with R.C. Herman, a postdoc from Princeton working at Johns Hopkins, and after calculating that there were one billion photons per nucleon, they used an early computer at the Bureau of Standards to extrapolate the Friedmann equations back in time. They discovered that the energy density of this radiation field must have initially considerably exceeded that of matter (as defined by $E = mc^2$). Radiation density is the amount of energy in a given volume of space, and it can be expressed as the temperature of the black body emitting the same energy. The observed 75/25 per cent hydrogen/helium ratio* implied that the temperature of the radiation must have been 10^9 K. Electromagnetic radiation readily interacts with electrically charged protons and electrons, so while the fireball was dominated by the radiation field the mean distance travelled by a photon was minuscule. As the Universe expanded, the radiation field would have 'cooled'. After some 300,000 years, its temperature would have fallen to 10^3 K, cool enough for electrons to associate with protons to produce neutral hydrogen atoms. As the radiation field 'decoupled' from the matter, the mean distance that a photon could travel was made near-infinite, and the Universe became transparent. In 1948

---

*   By way of clarification, the primordial helium abundance is about 10 per cent by number, and 25 per cent by mass.

Gamow announced the 'hot fireball' in a paper entitled 'The Evolution of the Universe'. In reporting their detailed analysis later in the year (in a similarly entitled paper) Alpher and Herman pointed out that as a result of the ongoing expansion, this super-hot radiation field would have been redshifted down to a temperature of about 5 K (the actual value depending upon the age of the Universe, which was not known precisely), placing its peak intensity in the microwave part of the radio spectrum. Although Gamow thought that this ought to be detectable as an isotropic background field, radar engineers assured him that there was no sufficiently sensitive detector, so he made no attempt to encourage radio astronomers to conduct a search. In 1952 Gamow published a popular account of the theory in a book, *The Creation of the Universe*, and sent a copy to Pope Pius XII, who wrote back saying that it was a confirmation of the *Genesis* account. With little prospect of making further progress with the theory, Gamow relocated to the University of Colorado at Boulder in 1956 to pursue other interests, and Alpher and Herman accepted jobs in industry.

THE HEAVY ELEMENTS

Meanwhile, the origin of the heavy elements had been solved. Eddington argued in 1926 that the Sun sustained itself by 'burning' hydrogen, and he estimated the absolute temperature of its core as 15 million degrees. In 1938 Bethe worked out a catalytic process involving carbon, nitrogen and oxygen (known as the CNO cycle) for transforming hydrogen to helium, but the Sun's core was not hot enough to drive this process. Working with Charles Critchfield, he later devised the 'proton–proton' reaction that powers stars like the Sun. After studying mathematics and theoretical physics at Cambridge, Fred Hoyle set out in 1945 (a year before Alpher started his study of primordial nucleosynthesis) to investigate whether the elements were made in stars, and then published 'The Synthesis of the Elements from Hydrogen' in 1946 and 'On the Formation of Heavy Elements in Stars' in 1947. Having studied physics at Ohio State University in 1933 and secured his doctorate in 1936 from Caltech, where he remained, W.A. Fowler decided in 1946 to measure the reaction rates to refine the theory. Bethe had not been able to explain how the carbon, nitrogen and oxygen that the CNO process used were produced. In 1952, in 'Nuclear Reactions in Stars without Hydrogen', E.E. Salpeter discovered that the obstacle at 5 nucleons (and indeed that at 8 nucleons) could be overcome in a helium-rich stellar core. Normally if two helium nuclei fused to form beryllium-8, this would decay as it is unstable. However, if the helium gas was sufficiently concentrated, there was a significant chance of the short-lived beryllium fusing with a third helium nucleus to create a carbon-12 nucleus. The helium density in the primordial fireball was too low for this, as indeed it is in the core of a solar-type star. The process can occur only in a star in which the hydrogen in the core has been converted to helium, which 'flashes' as the core is gravitationally compressed to initiate a new cycle of burning. In 1954–1955 Fowler joined Hoyle and the husband and wife team of Geoffrey and Margaret Burbidge at Cambridge, and they thrashed out the entire nucleosynthesis process for elements beyond helium. In 1957 the results were published in *Reviews of Modern Physics* in 'Synthesis of the Elements in Stars', a paper that became so well cited that it was often referred to by the initials of its authors as B^{2}FH. However, they could not account for the 25 per cent helium abundance, and the fact that Gamow's team *had* been able to explain this ratio had been overlooked because the primordial nucleosynthesis theory had been dismissed.

STEADY STATE

Thomas Gold and Hermann Bondi both fled Austria in the 1930s and settled in England. They met Hoyle while working on radar systems at the Admiralty during the Second World War. In the evenings, they discussed Gamow's form of Lemaître's 'cosmic egg', which they so despised that they sought an alternative explanation for the evident expansion of the Universe. After the war they all went to Cambridge. One day, Gold posed some key questions: "Why should all the matter of the Universe have been created at once, at the beginning of time? Why couldn't it have been created a little at a time, at a very slow rate, on an ongoing basis?" In 1928 J.H. Jeans had speculated that matter might be being continuously created, but Bondi, Gold and Hoyle developed this into a full theory. Although 'continuous creation' appeared to violate the conservation law stating that the total amount of energy in the Universe remained constant, so too did the creation of all that energy in a primordial fireball. Bondi suggested a refinement of the law, stating that the total energy of the *observable* Universe was constant, arguing that the energy introduced locally by continuous creation was balanced by the energy lost as the light from the most distant galaxies was redshifted to insignificance. It was a fine distinction, but it was plausible. The rate of creation was very low, only a few atoms of hydrogen per cubic kilometre per year. In the 1930s, E.A. Milne at Oxford had postulated the Cosmological Principle, which stated that the Universe must appear essentially the same to every observer, regardless of their position within it. Bondi, Gold and Hoyle extended this idea to time as well as to space, as the Perfect Cosmological Principle, which stated that the Universe *remains* essentially the same, and they proposed a 'steady-state' Universe which had always existed, was infinite in extent, and expanded without 'evolving'. Although galaxies were moving apart as space expanded, new ones were forming in between from the matter that was spontaneously created there. It was an ingenious alternative to the idea of a 'creation event'. In the division of labour, Bondi and Gold wrote 'The Steady-State Theory of the Expanding Universe', and Hoyle wrote 'A New Model of the Expanding Universe', both in the *Monthly Notices of the Royal Astronomical Society* in mid-1948 (which was a bumper year for cosmology papers). Although some theorists – particularly in Europe – found the idea appealing, others dismissed its basic premise.

In a series of BBC radio lectures in 1949 (published the following year in a slim volume entitled *The Nature of the Universe*) Hoyle derided the idea of a 'primeval atom' as "the 'big bang' theory of the Universe". Although the astronomers initially referred to Gamow's proposal as 'the radiation origin' in their formal publications, Hoyle's term caught the public imagination and the professionals soon followed suit. During the 1950s, therefore, the focus of cosmology was the search for evidence that would decide between all the competing theories.

THE RELIC OF THE FIREBALL

Upon being hired by the Bell Telephone Company at Holmdel in New Jersey in 1928, K.G. Jansky was assigned the job of identifying the various sources of 'noise' that impaired the recently introduced transatlantic radio-telephone service. He built a large rotating antenna on nearby Crawford Hill, and set to work. Having eliminated all known sources, he concluded in 1931 that there was a residual 'hiss' coming from the sky. By 1932 he had identified the source as lying in the direction of the centre of the galaxy. Although Jansky published his results in 1933 in the *Bell Systems Technical Journal* – which few, if any, astronomers read – *The New York Times* ran the discovery as its lead story on 5 May 1933: 'New radio waves traced to the centre of the Milky Way'. In 1937 Grote Reber, a radio engineer in Wheaton, Illinois, built a dish antenna on a steerable mount in his back yard in order to chart the radio sky at a wavelength of 2 metres. In 1938 he took his map to the nearby Yerkes Observatory, where Otto Struve, the director, noted that there was strong emission in the plane of the Milky Way. Furthermore, Struve realised that the dust that impeded the optical view was transparent to radio, and Reber had detected sources *at* the galactic centre. Other astronomers were sceptical but, with Struve's support, Reber published his results in 1942. After the Second World War, 'radio astronomy' was enthusiastically developed by A.C.B. Lovell at the University of Manchester, by J.S. Hey at the Royal Radar Establishment at Malvern, by Martin Ryle at Cambridge in England, by J.G. Bolton, J.L. Pawsey and B.Y. Mills in Australia, and by J.H. Oort, who in 1945 was appointed director of the Leiden Observatory in the Netherlands. In 1944 Oort had asked H.C. van de Hulst to calculate wavelengths at which hydrogen gas should be detectable in interstellar space. By orienting themselves with, or opposing, the weak ambient magnetic field, cold neutral hydrogen atoms adopt slightly different energy states. Furthermore, atoms that absorb a photon at a wavelength of 21 centimetres and adopt the higher state will eventually re-emit the photon when they resume the lower energy state. Any given atom might make such a transition only once in 10 million years, but hydrogen is so abundant that the result is continuous emission. After the war, Oort set out to construct a receiver to test this prediction, but in the summer of 1951 Felix Bloch at Stanford and E.M. Purcell and his graduate student, H.I. Ewen, at Harvard independently beat him to it. Radial velocity measurements of the 21-centimetre line in the 1950s confirmed the spiral structure of our galaxy.

In 1960 NASA launched the Echo satellite, a big inflatable sphere of aluminised-mylar to serve as a passive relay for transatlantic microwave communications. In order to receive signals that were bounced off the satellite, Bell (now part of AT&T) built a steerable horn antenna with a 20-foot-square aperture at Crawford Hill.

In 1960 the Bell Telephone Company built a steerable microwave horn antenna at Holmdel in New Jersey for use with early communications satellites.

During its calibration in 1961, E.A. Ohm noted that the antenna had an 'excess temperature' of about 3 K. Although this did not interfere with its satellite role, he reported it in the *Bell Systems Technical Journal*. In early 1962, Bell built a ground station at Andover in Maine to communicate over the Atlantic using the active-relay Telstar satellite. A matching station was being built in France, but its construction was behind schedule. After his graduation in physics from the City College of New York in 1954, A.A. Penzias joined the army. He later developed a maser (in effect an extremely sensitive microwave amplifier) while working for his doctorate at Columbia and undertook a fruitless search for 21-centimetre emission from putative intergalactic hydrogen. In 1961 he joined Bell and in 1962 was told to upgrade the Holmdel horn to serve as a backup station to demonstrate Telstar relays. By the time that the satellite was launched, however, the French station was ready and the horn, now surplus to requirements, was released to Penzias to be adapted for use in his maser research in radio astronomy. In 1963 R.W. Wilson was hired to help him. After graduating from Rice University in Houston in 1957, Wilson had secured his doctorate from Caltech in 1962, working with the Australian radio astronomer J.G. Bolton to map the 21-centimetre emission of hydrogen in the Milky Way. Penzias and Wilson were therefore kindred souls. As there was only one funded post, they job-shared and made up time on miscellaneous company tasks. For sensitive astronomical work the horn had to be finely calibrated and all sources of 'noise' eliminated. Upon rediscovering the hiss which Ohm had reported, they tried to track down its source but it was constant, irrespective of the direction in which they turned the antenna, irrespective of time of day, and indeed irrespective of the time of year.

After graduating from Princeton in 1938 R.H. Dicke gained his doctorate in 1941 from Rochester and spent the Second World War at the MIT Radiation Laboratory. While working on radar he developed a microwave radiometer. As an opportunistic test, while measuring how the atmosphere absorbed microwaves at wavelengths of 1, 1.25 and 1.5 centimetres, he searched for any celestial background. By finding nothing, he thereby set an upper limit of 20 K for the temperature of any such radiation field. A few months after this negative result appeared in *Physical Review* in 1946, the same journal published 'The Expanding Universe and the Origin of the Elements' in which Gamow predicted a primordial fireball. After the war, Dicke returned to Princeton. In 1963, he developed an interest in cosmology, specifically the so-called 'oscillating' Universe which, because it was 'closed', would eventually cease to expand and would collapse in on itself and – it was speculated – 'bounce'. Dicke was drawn to this model because the repeating cycles allowed the Universe (in the largest sense of the word) to be everlasting without being steady state. He was particularly interested in the question of whether it would be possible for any matter to survive the resulting 'Big Crunch' (a play on Hoyle's term for the creation event) and pass through from one cycle to the next. Perhaps the helium abundance had been built up in this way, he mused. After graduating in physics from the University of Manitoba in Canada, P.J.E. Peebles went to Princeton for his doctorate under Dicke, then remained there on a fellowship. In the autumn of 1964 Dicke asked Peebles to calculate the conditions in the final stage of the collapse. Unaware of the work by Gamow's team, Peebles trod the same ground and calculated that the temperature of the radiation would be of the order of 10^9 K, with its peak of emission in the gamma-ray range. As such radiation would dissociate nuclei into their constituent nucleons, no coherent matter could pass from one cycle to the next. Peebles further noted that the primordial radiation (from either the origin event or the start of the current cycle of the oscillating Universe, whichever was appropriate) ought to be detectable today as an isotropic field with a black body temperature of about 10 K. On 19 February 1965, Peebles presented a summary of his results in a seminar at Johns Hopkins University in Baltimore, saying: (a) if the early Universe had not been dominated by the radiation of a fireball, all of the superdense hydrogen would have been fused into helium; (b) the helium abundance was a measure of the energy density at the time that the fireball ceased to be dominated by radiation; and (c) the 'relic' of this field should be visible as an isotropic background. Kenneth Turner, a radio astronomer at the Carnegie Institution in Washington, DC, attended the talk and later related the prediction of a 10 K field to B.F. Burke of MIT. Meanwhile, Dicke asked P.G. Roll and D.T. Wilkinson to verify this. Operating at 3.2 centimetres, their radiometer was much smaller than the Holmdel horn – its aperture was only a foot across – so it was put on the roof of the laboratory. In early March, Peebles wrote his conclusions in 'Cosmology, Cosmic Black Body Radiation and the Cosmic Helium Abundance' and sent it to *Physical Review*, but the referee rejected it because simply referring to the review paper 'The Origin and Abundance Distribution of the Elements' written by Alpher and Herman towards the conclusion of their research did not fully recognise the work by Gamow's team. Although Peebles expanded the list of references to include 'On Relativistic Cosmogony' (an important paper by Gamow) and 'Physical Conditions in the Initial Stages of the

Expanding Universe' (by Alpher, Follin and Herman) his article was again rejected for failing to acknowledge the 1949 paper 'Remarks on the Evolution of the Expanding Universe' in which Alpher and Herman predicted a background at 5 K. It was pointless for Peebles to rectify this omission as it would render his own prediction sterile, so his paper fell by the wayside. A more general paper on 'Gravitation and Space Science', written at this time with Dicke as the lead author, was accepted by *Space Science Review*. Between its submission and its appearance, a crucial discovery was made.

In March 1965, having tried everything else they could think of in their efforts to eliminate the noise in their antenna, Penzias and Wilson scraped off the accumulated bird droppings, but most of the hiss remained. Concluding, reluctantly, that the noise would be a limiting factor in the use of the antenna for astronomical research, they were about to "bury it" in a footnote in their report when Penzias received a telephone call from his friend B.F. Burke. He had encountered Burke while returning from a conference in Montreal, Canada, in December 1964 and told him of his frustration in tracking down the mysterious hiss. Burke suggested that Penzias should contact Princeton, as Peebles had presented a seminar in which he predicted the existence of a microwave background. Dicke sent Penzias a copy of Peebles's draft and, after reading it, Penzias invited Dicke to drive over (it was only 50 kilometres) to inspect the signal. Although irked not to have been the first to make the discovery, Dicke's team were delighted with the Holmdel data. The radiation temperature was slightly cooler than predicted, at only 3 K, but its isotropy was the tell tale. Even though the Holmdel horn was the world's most sensitive microwave receiver, when Dicke set out to test Peebles's prediction it had not occurred to him to employ it because it was a communications relay rather than a radio telescope. The *Astrophysical Journal* agreed to publish companion papers. On 13 May, Penzias and Wilson submitted 'A Measurement of Excess Antenna Temperature at 4080 Mc/s'. It was terse (barely 600 words) and strictly factual, venturing only that "a possible explanation for the observed excess noise temperature is the one given by Dicke, Peebles, Roll and Wilkinson in a companion letter in this issue". Although a black body curve could be drawn through the Holmdel measurement at 7 centimetres, measurements at a range of wavelengths would be needed to prove the case. Nevertheless, Dicke's team were vigorous in promoting their interpretation, and boldly entitled their paper 'Cosmic Black Body Radiation'. Having attended Hoyle's lectures on steady-state cosmology at Caltech and found that theory appealing, Wilson was initially sceptical. Penzias was simply delighted to have someone explain the mysterious excess. Before the papers appeared in July, Peebles reworked his calculations with the temperature set at 3 K in order to refine his estimate for the amount of helium that would have been made in the fireball. When giving a seminar to the American Physical Society he prompted howls of laughter by presenting a viewgraph of a black body curve passing through a single data point. The science writer Walter Sullivan broke the story on 21 May, when *The New York Times* ran it as the lead item, above the fold on page one with the banner "Signals imply a 'Big Bang' Universe". Continuing: "Scientists at the Bell Telephone Laboratories have observed what a group at Princeton University believes may be the remnants of an explosion that gave birth to the Universe." Lemaître (who lived until 1966) was undoubtedly delighted to read that his cosmology had been proved correct. Gamow was disappointed that neither paper had acknowledged his team's work. Despite Peebles's efforts to review prior research, his team's paper cited only the 1953 paper 'Physical Conditions in the Initial Stages of the Expanding Universe'; no reference was made to the earlier papers predicting a 10 K background. Evidently the referees of the *Astrophysical Journal*, noting the companion paper confirming the prediction, passed the new paper with less than adequate citations.

MISSED OPPORTUNITIES

Although microwave detectors were insufficiently sensitive to detect such a faint background signal in the 1950s, the fact that interstellar space is bathed in radiation *had* been inferred by optical astronomers. Improvements in spectroscopy during the 1930s revealed absorption lines of the cyanogen radical. As such a delicate molecule could not exist in a stellar envelope, it was realised that the lines were superimposed upon the spectrum as the light travelled through clouds of gas in interstellar space. In 1940, W.S. Adams of the Mount Wilson Observatory made high-resolution spectra of cyanogen in an 'excited' rotational state. Andrew McKellar of the Dominion Astrophysical Observatory in British Columbia, Canada, interpreted this as meaning that the gas in the intervening cloud was bathed in radiation that had a temperature in the range 0.8 to 2.7 K. When a second line was predicted and confirmed, he was able to refine this value to 2.3 K. As this item of minutiae was included in *Spectra of Diatomic Molecules*, the standard reference work for spectroscopists, Gamow may well have become aware of it. Nevertheless, he did not make the connection when Alpher and Herman predicted a background radiation field at 5 K. While mapping emission from the Milky Way at a wavelength of 33 centimetres in 1955, the French radio astronomer Émile Le Roux set an upper limit for the temperature of any isotropic background as 3 (± 2) K. At various times Gamow had calculated temperatures ranging as high as 50 K. In 1956, shortly before Gamow went to Colorado, Hoyle paid him a visit and noted that the evidence not only ruled out a temperature towards the upper end of Gamow's range, but also seriously questioned the 5 K value. The field was first detected in 1957, when T.A. Shmaonov in the Ukraine found that a horn antenna had an excess temperature of 4 (± 3) K, but when radio astronomer Yuri Pariskij said that an isotropic signal was unlikely to be of celestial origin he rejected it as instrumental error. In 1960, while independently following in McKellar's footsteps, G.B. Field at Princeton concluded that cyanogen lines implied that interstellar space was pervaded by a radiation field at about 3 K, but Lyman Spitzer dissuaded him from publishing. William Rose of the Naval Research Laboratory in Washington, DC, noted it in 1962, but as he did not realise the significance of the observation he did not publish. In 1963 A.G. Doroshkevich and I.D. Novikov in Russia, having read Ohm's report of an excess in the Holmdel horn at 7 centimetres, recommended that tests be conducted at longer wavelengths (where galactic emission was weaker) in order to test Gamow's prediction. They even ventured that the Holmdel horn was the best instrument for such a search. Ya.B. Zel'dovich began his career as a laboratory assistant in Minsk's Institute of Useful Ores, but was transferred to the Physical Technical Institute in Leningrad in part exchange for a vacuum pump, and thus changed his field from chemistry to physics in which, five years later, he gained his doctorate. After the Great Patriotic War, he was a leading member of the team that developed the Soviet atom bomb. In 1963 he investigated Gamow's theory, and upon concluding that the background should have a temperature of 20 K, interpreted Ohm's measurement as *disproving* the hot fireball.

In the early 1960s, Hoyle, aware that stellar nucleosynthesis could not account for the observed helium abundance, joined with Roger Tayler to calculate how much helium would have been made in the hot Big Bang (were, perish the thought, such an event to have occurred) and they concluded that fireball nucleosynthesis would have produced *at least* 14 per cent helium. In their 1964 paper 'The Mystery of the Cosmic Helium Abundance', they narrowly missed predicting the background field. Hoyle's Caltech collaborator, W.A. Fowler, then asked R.V. Wagoner, a young Stanford physicist who had just joined his team, to calculate the heavier elements that would emerge from the fireball. In 1966, in 'Nucleosynthesis in the Early Stages of an Expanding Universe', they confirmed that the process would stall after helium, whose abundance would be about 25 per cent. It is ironic, therefore, that Hoyle, one of the most vocal critics of a primordial fireball, should have co-authored the seminal work that convinced most cosmologists to take the idea seriously! The result of all this work was what came to be known as the 'standard model' of the Big Bang.

FOLLOWING THE CURVE

Once Holmdel's 7.35-centimetre result became known, other astronomers turned their antennas to the sky to confirm the field's black body character by measuring it at a range of wavelengths. Measurements at 3.2 centimetres by Dicke's team and at 21 centimetres by Penzias and Wilson (once they had reconfigured the Holmdel horn) were consistent. Other teams extended the coverage to 75 centimetres, in each case following the 3 K curve. However, all these measurements were longward of the predicted turnover in the spectrum at about 2 millimetres, and the crucial test would be whether the spectrum declined sharply shortward of this. The atmosphere impeded ground-based submillimetre studies, so in 1968 detectors were launched on two sounding rockets by Martin Harwit and James Houck at Cornell and Kandish Shivanandan of the Naval Research Laboratory. The signal short of the turnover was much stronger than could be accounted for by a 3 K field, but there was a possibility that the instruments had suffered some sort of interference. In 1971 a measurement at about 1.5 millimetres was made by an MIT team using a balloon-borne instrument and was generally consistent with the 3 K curve. In 1974 a Berkeley balloon-borne detector indicated considerable excess radiation, but the uncertainties inherent in the data were significant and it was impossible to state definitively that the results were inconsistent with the prediction.

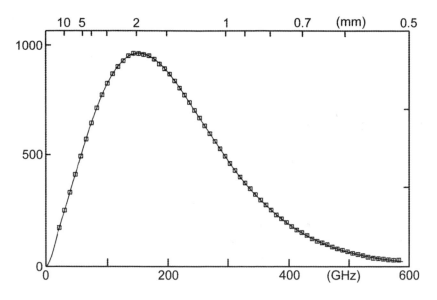

In November 1989 NASA launched the Cosmic Background Explorer (COBE) with an instrument to measure at 67 wavelengths from 0.5 to 10 millimetres. The results confirmed the black body spectrum at a temperature of 2.726 ($\pm$0.01) K – the error bars on the individual measurements were so small that the points precisely traced out the predicted curve. Courtesy of J.C. Mather, COBE/FIRAS Team Leader.

Despite the ambiguous data from the early rocket- and balloon-borne detectors, the tide had turned in favour of the Big Bang. Indeed, at an international symposium entitled 'Confrontation of Cosmological Theories with Observational Data' in Cracow in 1973 to celebrate the 500th anniversary of the birth of Nicolaus Copernicus, almost every paper was in favour of the Big Bang. It was belatedly realised that Einstein's working assumption that the Universe was perfectly smooth and isotropic was not the gross oversimplification that he had believed. Inconclusive measurements short of the turnover continued through the 1980s. In 1987 Berkeley's detector, now modified to measure at three submillimetre wavelengths, was launched on a Japanese sounding rocket and again reported an excess above the black body curve. Another instrument was sent up in July 1989 as a check, but a wiring fault produced electrical interference that compromised the data. It was a frustrating time for experimenters and theorists alike, but the importance of the issue had prompted the development of a satellite to provide a definitive answer. In November 1989 NASA launched the Cosmic Background Explorer (COBE) with instruments to measure the energy at 67 wavelengths from 0.5 to 10 millimetres; to determine the extent to which the field was isotropic; and to investigate other sources contributing to any excess short of the turnover. The first 10 minutes worth of data confirmed the black body spectrum – the error bars on the individual measurements were so small that the points precisely traced out the predicted curve. When J.C. Mather of Berkeley presented the preliminary data to a meeting of the American Astronomical

Society in Arlington, Virginia, in January 1990, the audience rose for a standing ovation – a very rare occurrence at scientific gatherings. The final report put the temperature of the black body at 2.726 ($\pm$0.01) K. After several years and many millions of individual measurements for an all-sky temperature map, COBE revealed the background to be isotropic to 1 part in 100,000. In fact, it was so smooth that the maximum departure from the mean was a mere 30-millionths of a degree. If the distribution of matter was *perfectly* smooth at the time that the radiation decoupled, the interaction of photons with matter would have been homogeneous and produced a *completely uniform* background. COBE established that matter was *not* uniformly distributed at that time; there was already a variation in density. "What we have found", pointed out G.F Smoot, leading the team, "is evidence for the birth of the Universe." His colleague, Joseph Silk, described it as "the 'missing link' of cosmology". It was from such 'wrinkles' that the large-scale structures that we see today developed.

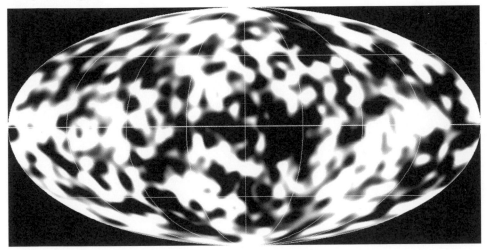

An instrument on the Cosmic Background Explorer (COBE) sought fluctuations in the intensity of the microwave background across the sky, and found that although the field was smooth to within 1 part in 100,000, there were primordial 'ripples'. Courtesy of G.F Smoot, COBE/DMR Team Leader.

9

Probing the furthest reaches

A MATTER OF SCALE

After being awarded his doctorate from Harvard in 1937, J.L. Greenstein spent a few years at the Yerkes Observatory before joining the faculty at the University of Chicago. In 1948 he was recruited by Caltech to establish its graduate astronomy school. A.R. Sandage, a physics graduate from the University of Illinois, was one of the first intake. Acknowledging that he could no longer sustain a vigorous observing campaign after suffering a heart attack in 1950, Hubble began to send Sandage up to the 200-inch to expose plates. By the time of Hubble's death in 1953, Sandage had taken on the mission of extending the redshift–distance relationship.

Humason worked with N.U. Mayall of the Lick Observatory in the compilation of redshifts of 850 galaxies for the most comprehensive survey to date (extending the recessional velocity to 100,000 kilometres per second) which was published in 1956. Sandage, using the 200-inch, had started by searching for cepheids in nearby galaxies in order to confirm the distances inferred by Hubble who, being unable to observe cepheids with the 100-inch, had relied upon an extrapolation of the brightest stars. Sandage found that many of the objects that Hubble had presumed to be stars were actually vast clouds of glowing hydrogen, and being more luminous were much more remote. After rectifying this flaw, Sandage further reduced Hubble's constant to 55 kilometres per second per million lightyears, and increased the extrapolated age of the Universe to 5.5 billion years. In 1953 Clair Patterson, a young geochemist at Caltech, had used a meteorite to calibrate a method of measuring the age of the Earth, which was 4.5 billion years. While 5.5 billion years was 1 billion years more than the age of the Solar System, it was still insufficient for the globular clusters. On this revised scale, the 1956 survey ranged out several billion lightyears, sufficiently deep to encompass 18 rich clusters. In each case, there was a giant elliptical (often located right in the centre) that far outshone its companions. Sandage found that plotting the apparent magnitude against redshift for these 'first ranked' ellipticals gave a straight line, implying that they were equally luminous and could be used as standard candles with which to probe even further. It was an important discovery in its own right, as it suggested a maximum size for such galaxies. Having recalibrated the distance scale, in 1961 Sandage reduced the value of Hubble's constant to 50 kilometres per second per million lightyears. However, the apparent brightnesses of the globular clusters in galaxies in the Virgo cluster led him to suspect that it might be much less.

After Baade returned to Germany in 1958, and Humason retired, Sandage ruefully reflected that "the entire nebular department consisted of me". With ready access to the 200-inch, he was accumulating plates faster than he could analyse them. When he met G.A. Tammann of the University of Basel in Switzerland at a summer school in Italy in 1962, he invited him to join in the analysis. Since 1954 Sandage had been seeking cepheids in M81, a giant spiral that dominated a small cluster in Ursa Major which Hubble had deemed to be not too far off. Sandage had identified 40 candidates for variable stars, but had been unable to classify them. In effect, he was limited to the longer period cepheids, which were more luminous, but it typically required at least 30 plates over a 10-year interval to determine their periods. Upon arriving at Caltech in early 1963, Tammann had no success with M81, so they turned their attention to NGC2403, which was another spiral in that cluster. After a year and a half of painstaking analysis Tammann had identified a dozen cepheids, but their light curves were sketchy because they were visible only near their peaks, disappearing into the murky background in between. Nevertheless, in 1967 they announced that the distance to NGC2403 was some 10.6 million lightyears, which reinforced Sandage's suspicion that the value of Hubble's constant was still overestimated.

The next major target was M101, the brightest spiral of another group of galaxies in Ursa Major. Unfortunately, they could neither find any cepheids nor identify any of the supergiants for which they had a calibration. These negative results nevertheless indicated that this galaxy was remote. They could see some glowing hydrogen clouds, but there was no calibration for a spiral of that size. Turning their attention to the irregular galaxies in the cluster of a size for which they did have suitable calibrations, they found a distance of 22 million lightyears, which was much more than expected. On the reasonable assumption that M101 was more or less at the same distance, they measured its overall luminosity in order to use giant spirals of this class as they had been using 'first ranked' giant ellipticals. In 1975 Sandage dramatically cut the value of Hubble's constant to 17 kilometres per second per million lightyears. However, he suspected that it might fall to 15 kilometres per second per million lightyears, which would increase the maximum age of the Universe to 20 billion years, which was more than sufficient to accommodate the globular clusters.

MYSTERIOUS QUASARS

After graduating in physics from Oxford in 1939 Martin Ryle worked on radar during the Second World War, then went to Cambridge where he started a group

This picture of the spiral galaxy M81 in the constellation Ursa Major was taken in 1975 with the 4-metre Mayall Telescope at the Kitt Peak National Observatory. Courtesy of NOAO/AURA/NSF.

This picture of the giant spiral galaxy M101 in the constellation Ursa Major, dubbed the 'Pinwheel', was taken in 1975 with the 4-metre Mayall Telescope at the Kitt Peak National Observatory. Courtesy of NOAO/AURA/NSF.

This picture of NGC5128, a peculiar elliptical galaxy identified as the Centaurus A radio source, with a dark dust lane silhouetted against its nucleus, was taken with the 4-metre Blanco Telescope of the Cerro Tololo Interamerican Observatory in Chile. Courtesy of NOAO/AURA/NSF.

that pursued radio astronomy. In 1945 J.S. Hey inspected a strong radio source which Grote Reber had charted in Cygnus. Upon observing that its intensity fluctuated on a time-scale of minutes, he inferred that its source must be physically small. He named it Cygnus A. In 1947 J.G. Bolton erected an antenna on a cliff overlooking Sydney Harbour in order that a discrete source rising over the horizon could be observed both directly in the sky and by its reflection off the sea. This interferometry gave much finer angular resolution than the antenna could achieve alone. Using this technique, he identified Taurus A with the Crab nebula, and tentatively associated Virgo A with M87 (a giant elliptical with an intriguing jet projecting from it) and Centaurus A with NGC5128 (an unusual galaxy with a dark dust lane silhouetted against its nucleus). He was, however, too far south to study Cygnus A.

This short-exposure of the giant elliptical galaxy M87 was taken to document the prominent jet, whose blue light contrasts with the red starlight. The synchrotron emission is strongly polarised.

In 1948 Ryle built a two-element interferometer utilising a field of dipole aerials with a separation of 1,000 feet to scan the sky overhead as the Earth rotated. He soon realised that the discrete sources fluctuated as a result of scintillation induced as their radio waves transited the Earth's ionosphere, much as the atmosphere makes stars 'twinkle'. As this meant only that they subtended small angles, they were not necessarily physically compact: they could be much larger and further away. In 1950 Ryle published the *Cambridge Catalogue* of radio sources. The discrete sources were presumed to be 'radio stars', although none had been identified with a specific star. At a conference in 1951 (by which time 50 had been discovered) Fred Hoyle and Thomas Gold noted that, as their distribution was not concentrated towards the plane of the Milky Way, they were probably extragalactic. Introducing the term 'radio galaxies', they pointed out that several sources could reasonably be associated with known galaxies. Later in the year, F.G. Smith at Cambridge pinned down Cygnus A sufficiently to allow an optical search for its counterpart. He airmailed the

"I have never seen anything like it before," observed W.H.W. Baade upon seeing this double-lobe feature on a 'deep' plate which he had taken through the 200-inch Hale Telescope at the position of the Cygnus A radio source in 1951.

information to Baade, who photographed the location through the 200-inch and found a rich cluster of galaxies, at the centre of which was a peculiar object of 18th magnitude that appeared to have a distorted wispy structure and a double nucleus. "I have never seen anything like it before", he wrote to Smith.

Having recently co-authored a paper with Lyman Spitzer speculating upon colliding galaxies, Baade concluded that this was such an event. Minkowski secured a spectrum in which strong emission lines suggested the presence of a hot plasma. Although it was one of the brightest radio sources in the sky, the redshift of 6 per cent meant that it was almost 1 billion lightyears distant. Could a collision produce a superheated plasma? Could hot gas generate so much radio emission? In 1953 R.C. Jennison in England discovered that the discrete source was situated between two lobes of radio emission which, together, formed a vast dumbbell shape. Whereas the discrete source was only 10 seconds of arc across, the pair of lobes spanned fully 3 minutes of arc. Something in the galaxy was firing material 1 million lightyears into space. The vast scale of the radio lobes implied that some *sustained* process had been pumping out material for a considerable time. In an attempt to determine the speed of collision, the failure to discern a difference in the radial velocities of what were presumed to be the two separate nuclei served only to undermine the hypothesis. V.A. Ambartsumian in Russia argued in 1955 that such unusual galaxies were undergoing catastrophic outbursts. In 1958 Geoffrey Burbidge pointed out that the radio output indicated a 'non-thermal' process such as synchrotron radiation from electrons travelling in an intense magnetic field, which in turn implied that there was a 'powerhouse' of some sort in the core of the galaxy.

While A.C.B Lovell at Manchester University sought funding to build a 250-foot dish antenna, Ryle continued to pioneer interferometry – what it lacked in raw sensitivity was compensated by its ability to resolve fine detail. Although the four-element interferometer commissioned in 1955 was manually steered, it was effective and in 1958 the *Third Cambridge Catalogue* (known as '3C') was issued. As the number of sources increased, the fact that progressively fainter ones were more populous was an indication that the Universe had *evolved*, which was contrary to the prediction of the steady-state theory's Perfect Cosmological Principle.

When the super-wide-field 48-inch Schmidt telescope-camera entered service on Mount Palomar in 1949, Minkowski supervised the massive task of compiling the National Geographic Society's Palomar Observatory Sky Survey. In the spring of 1960, Minkowski, one of the first optical astronomers to realise the significance of radio astronomy, received an unpublished position for 3C295 in Boötes. On his final

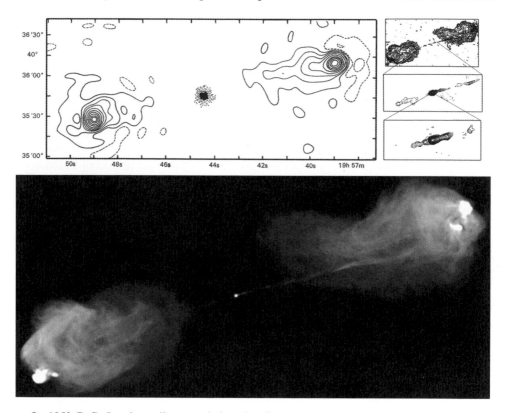

In 1953 R.C. Jennison discovered that the discrete source known as Cygnus A was situated between two vast lobes of radio emission, indicating that something in this unusual galaxy was firing material a million lightyears into space. As the resolution of radio telescopes was improved, it was possible to investigate the structure, and with the advent of computer processing it became possible to represent such maps as visual images. A view at a wavelength of 6 centimetres by the Very Large Array in Socorro in New Mexico shows the narrow jets of relativistic electrons emerging from the compact core where the stars are located, and 'hot spots' formed by the shock waves where the jets impinge upon the vast lobes of dispersing material. For further details, see either 'Cygnus A', C.L. Carilli and P.D. Barthel, *Astronomy and Astrophysics Review*, vol. 7, pp. 1–54, 1996, or *Cygnus A: Study of a Radio Galaxy*, C.L. Carilli and D.E. Harris (Eds), Cambridge University Press, 1996.

session on the 200-inch before retiring, he spent two nights integrating its spectrum. It was a peculiar galaxy with emission lines similar to those of Cygnus A. However, 3C295's 45 per cent redshift indicated that it was 7 billion lightyears away, making it the most distant object known. Minkowski had not been able to see anything of it through the prime focus, he had merely placed the spectroscope's slit on the radio position and guided on an offset field star. A deeper plate showed the same double nucleus as Cygnus A, suggesting a morphological similarity. This raised the tantalising possibility that there might be more objects of this type.

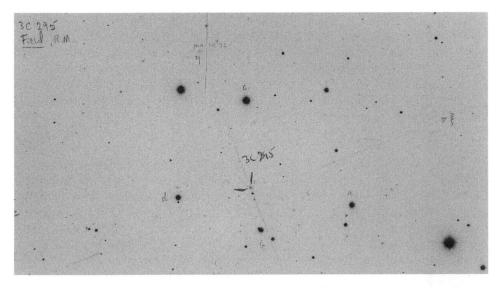

The Palomar Observatory Sky Survey negative which R.L.B Minkowski marked up in his search for the optical counterpart of 3C295 in Boötes in 1960.

T.A. Matthews was a radio astronomer with a Harvard doctorate. On moving to Caltech he helped to build a two-element radio interferometer at Owens Valley, east of the Sierra Madres. He had been given an unpublished list of refined positions for some of the 3C sources by a friend using Jodrell Bank's 250-foot dish and, noticing that 3C295 was the most point-like entry, he selected those that most resembled it and requested Sandage to examine 3C48 in Triangulum using the 200-inch. Sandage took multicolour photometry of the innocuous star-like object of 16th magnitude in order to classify it on the Hertzsprung–Russell diagram, and found it to be unusual. As he was not used to such *bright* objects, it took several attempts to get a spectrum that was not saturated. With a very strong ultraviolet excess and several mysterious emission lines, it was "the weirdest" that he had ever seen.

Cyril Hazard monitored an occultation of 3C273 in Virgo by the Moon in 1962 using the 210-foot dish of the Parkes telescope in Australia and not only refined its position sufficiently for an optical search but also discovered that it was *two* sources separated by 20 seconds of arc. Sandage found a 13th-magnitude star with a jet; one of the sources coincided with the star, and the other with the jet. It was reminiscent of M87, but what kind of *star* could produce a jet?

Maarten Schmidt was awarded his doctorate in astronomy by the University of Leiden in the Netherlands in 1956. Upon being recruited by Caltech two years later he worked with Minkowski in seeking optical counterparts of discrete radio sources. In May 1962 he took a spectrum of 3C286, which had only one emission line. Over the ensuing months he found 3C147 to have a few lines, but 3C196 to be featureless. On 27 December he took a spectrum of 3C273, which generally resembled 3C48 but with more lines. On 5 February 1963, in a moment of epiphany, he noted that some of the spectral lines were properly spaced for the hydrogen Balmer series, but with a 16 per cent redshift. This implied a velocity of 47,000 kilometres per second which, if interpreted in terms of Hubble's redshift–distance relationship, placed it *2 billion* lightyears away. If it was as far away as it seemed, then its jet had to be enormously long. For it to appear as a 13th-magnitude star, it had to be 100 times more luminous than any of the radio galaxies yet identified, the most powerful of which were the 'first ranked' members of their clusters.

Seeing Greenstein sauntering past the door of his office, Schmidt called him in and explained his discovery. As a specialist in stellar evolution, Greenstein had just sent off a paper describing 3C48 as the relic of a supernova. In fact, it had a redshift of 37 per cent. At a distance of 4 billion lightyears, it was more distant than all but a few galaxies and, apart from Minkowski's 3C295, was the most remote object yet measured. After a few of shots of whiskey, Schmidt checked his logic to ensure that he had not made a silly mistake, and Greenstein called the editor of the *Astrophysical Journal* to request that his paper be killed. That night Schmidt told his wife that he had discovered "something wonderful", and the next day he sent a one-page letter to *Nature* mundanely entitled '3C273: A Star-like Object with a Large Redshift', which was immediately followed by Greenstein and Matthews's 'Redshift of the Unusual Radio Source 3C48'. H.Y. Chiu introduced the term 'quasi-stellar radio sources', which was soon truncated to 'quasar'. It had not previously occurred to anyone to interpret the emission lines in terms of redshift because only extragalactic objects had redshifts. It had been "a mental block", admitted Sandage, astonished that he had not solved the mystery himself.

Noting that Cygnus A and Minkowski's 3C295 fitted his redshift–brightness plot, Sandage inferred that they were comparable to the 'first ranked' giant ellipticals used as standard candles. The prospect of using the highly redshifted quasars to probe the edge of the Universe was tantalising. The pace of quasar discovery was limited by the refinement of the radio positions, so in 1964 Sandage asked Ryle to send him the best 3C positions and then, assisted by graduate student Phillippe Veron, he began an optical search using a two-colour analysis to identify the 'stars' in the box with a strong ultraviolet excess. For such relatively bright objects he did not need the great light grasp of the 200-inch, so he utilised the 100-inch, which was actually more appropriate for a search as it offered a broader field of view. To his surprise, there were often several objects with an ultraviolet excess in a box, so he selected the one nearest the centre and confirmed it to be a quasar. 3C9 in Pisces (which was on Matthews's list) was found to have a redshift of 200 per cent. Nevertheless, it was

This picture of the radio source 3C273 in Virgo was taken with the 4-metre Mayall Telescope at the Kitt Peak National Observatory. Courtesy of NOAO/AURA/NSF. By recognising three emission lines of hydrogen in its spectrum Maarten Schmidt was able to calculate a redshift of 16 per cent.

still intensely blue because the Lyman-alpha line of hydrogen had been shifted down into the visible spectrum. In early 1965 Sandage, playing a hunch, examined several of the *other* objects in one of the 3C boxes and found that *they too* had high redshifts. They had characteristic quasar spectra, but were *not* emitting radio energy. In fact, these 'quiet' ones were found to be much more numerous. Sandage's hope that quasars might serve as standard candles was dashed when it was realised that they were highly variable, and so he lost interest. Schmidt, however, was smitten: "The quasars totally changed our insight and outlook upon the future study of the Universe". One hundred and fifty quasars had been found by the end of the 1960s, and of the two-thirds whose spectra had been investigated, all were *more* heavily redshifted than 3C273, which turned out to be the *nearest*!

After listing quasars in their star catalogues, optical astronomers would probably have remained ignorant of their cosmological distances if radio astronomers had not drawn attention to them. On the other hand, since they looked like stars, a search of the archives revealed that they were variable, and monitoring showed both radio and visual variability by up to 100 per cent over intervals as short as a week, indicating that their active regions spanned no more than 200 billion kilometres (a 'lightweek'). A doubling in brightness and a return to 'normal' by 3C446 in Aquarius within a 24-hour period indicated that the active region was a few 'lighthours' across. How could a volume smaller than the Solar System produce the output of 10 to 100 galaxies? A supernova can outshine its host galaxy, but it is necessarily a momentary event. Not only was the powerhouse of a quasar even more energetic, its output was sustained for billions of years. The hydrogen fusion process that powers the Sun converts only 0.7 per cent of the mass involved in a reaction into energy. An early suggestion was that quasars converted mass directly into energy with 100 per cent efficiency by $E = mc^2$, but what physical situation would prompt this? Another idea involved a novel concept. When Chandrasekhar found in 1930 that there was a maximum mass for a white dwarf, he presumed that a larger star would suffer runaway collapse. In 1934 Baade and Zwicky speculated that if a star that was too massive to become a white dwarf collapsed, the electrons would be forced into atomic nuclei, where they would combine with the protons to form a neutron gas. Like electrons, neutrons are fermions subject to Pauli's Exclusion Principle. In 1939 J.R. Oppenheimer and G.M. Volkoff, one of his graduate students, calculated how 'degenerate neutron pressure' would resist further gravitational collapse to create a 'neutron star'. Harlan Snyder, another of Oppenheimer's students, discovered that if the star exceeded about 3 solar masses, this pressure would be unable to inhibit Chandrasekhar's runaway collapse. Actually, this was not a new idea. In 1796 the French mathematician P.S. Laplace had realised that if a collapsing cloud of interstellar gas was sufficiently massive it would form a '*corps obscur*' whose tremendous gravitational field would prevent light from escaping. Prior to the 1916 publication of Einstein's paper on general relativity, Karl Schwarzschild in Russia had been thinking about whether the Universe could be described by non-Euclidean geometry. Whereas Einstein and de Sitter had pondered the cosmological implications, Schwarzschild mused on the prediction that light would be bent when passing close to a gravitating mass. He realised that if a star had a critical ratio of mass to radius it would *warp* space so severely that not even light would be able to escape. This became known as the 'Schwarzschild radius'. Although the term 'dark star' was coined, the fact that the masses of stars were evidently not confined within such small volumes meant it was dismissed as lacking physical significance. The properties of collapsed objects were later investigated in some detail, but it was believed that, in practice, the process of collapse would be disrupted by turbulence, preventing the *singularity* from forming and giving rise to an irregular explosion. In 1965 Roger Penrose at the University of London proved that a runaway collapse *must* end in a singularity. In 1964 E.E. Salpeter at Cornell pointed out that the angular momentum of gas attracted by a singularity's gravitational field would cause it to spiral in and form an encircling 'accretion disk', and that intense frictional heating would cause the gas to radiate X-rays. In 1965 Ya.B. Zel'dovich and O.H. Gusneyov in Russia reached this conclusion independently, and pointed out that if one star in a binary system formed a singularity, the X-rays from the gas drawn from its companion ought to be detectable. At a conference in New York in 1967, J.A. Wheeler of Princeton coined the term 'black hole', chosing this name because the singularity was a veritable 'hole' in the fabric of space. The critical radius was labelled the 'event horizon' because no events beyond it were visible. In contrast to the star that Laplace had considered, the event horizon is not a physical surface. Penrose opined that nature imposed the event horizon as an act of 'cosmic censorship' in order to hide the *flaw* from the rest of the Universe.

In 1965, a sounding rocket with an X-ray detector saw a strong discrete source in Cygnus, and it was duly named Cygnus X-1. In 1970 the Uhuru satellite was launched into orbit for an all-sky survey at X-ray wavelengths. Cygnus X-1 was soon observed to be 'flickering' on a millisecond time-scale which (by speed-of-light constraints) meant that the source had to be no more than 300 kilometres across. Once radio telescopes had refined its position, optical attention soon focused on HDE226868, a hot blue star of about 30 solar masses. Spectroscopic observations by C.T. Bolt of the University of Toronto showed that the X-ray source was orbiting this star with a 5.6-day period, and was 5 to 8 solar masses (this range reflecting uncertainty as to the inclination of the orbit to our line of sight). This was too much for a neutron star. The diameter of an event horizon is a linear function of mass – that of a 5-solar-mass black hole is 30 kilometres in diameter and that of a 10-solar-mass black hole is 60 kilometres across. By the early 1970s, therefore, it had been shown that stellar-mass black holes really existed. As a result, research took off. In 1974 Donald Lynden-Bell and M.J. Rees at Cambridge argued that quasars were accretion disks around 'supermassive' black holes in the cores of galaxies. Accretion disk processes liberate energy with about 10 per cent efficiency, which is considerably greater than by fusion. Nevertheless, quasars would have to convert between 0.02 and 20 solar masses per year to account for the observed range of luminosities and – depending upon the rate, with the more voracious ones being most luminous and attaining the largest masses – would grow to between 10^5 and 10^8 solar masses over 10 billion years of activity. In the most luminous cases, the active region would have to be gravitationally bound by *at least* 10^8 solar masses, otherwise the radiation pressure would disperse the accretion disk. Their colleague Roger Blandford then

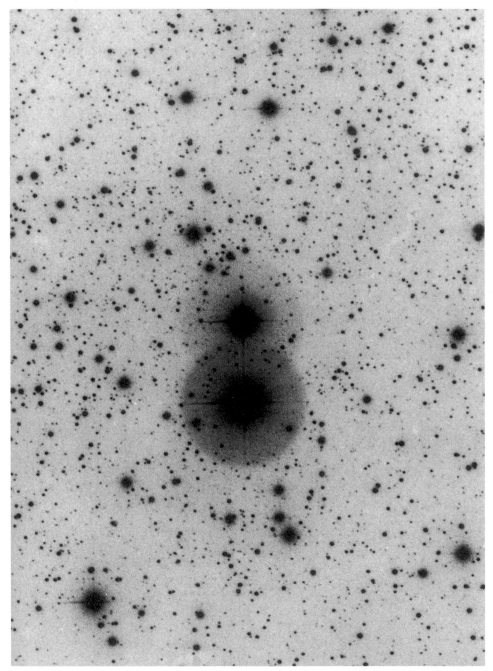

Once the position of Cygnus X-1 had been refined, optical attention focused on the blue supergiant HDE226868 (the brightest star on this negative taken with the wide-field 1-metre Swope Telescope at the Las Campanas Observatory on the Atacama desert of southern Chile) around which the X-ray source is orbiting with a 5.6-day period.

This artist's impression of the Cygnus X-1 system shows gas leaking from the blue supergiant star HDE226868, whose shape is distorted by its companion's presence, onto the accretion disk that circles a compact object of 5 to 8 solar masses which can only be a black hole. Artwork by L. Cohen of the Griffith Observatory in Los Angeles.

realised that the infalling *ionised* gas would 'wind up' the ambient magnetic field to an incredible intensity, which in turn would 'beam' plasma out along the magnetic axis. If this was how the radio lobes of Cygnus A were formed, then there must be a supermassive black hole in its heart, in which case it must be a nearby quasar in a 'quiescent' state.

The late 1960s saw the development of the 'standard model' of the Big Bang. Although this predicted a 'cut off' in the population of galaxies at high redshift, this transition was beyond the most distant measured galaxies and was therefore not evident, but the quasars offered the prospect of probing the early Universe, perhaps even to a time before galaxies formed. By 1973 only a few of the 200 redshifts were in the 80 to 90 per cent category. While this was consistent with a cut off, the trend was ill defined, so in 1974 Maarten Schmidt and Richard Green started a search using the wide field of view of Zwicky's 18-inch Schmidt Telescope on Mount Palomar equipped with a filter to highlight the intensely blue star-like objects. In a paper in the *Astrophysical Journal* in 1982 they listed 94 new quasars. The statistics supported the idea that the voracious black holes shone most brilliantly and 'switched off' soonest, whereas the less luminous ones lasted much longer. Although their filter could not distinguish *deeply redshifted* objects whose emission peak was now in the middle of the visual spectrum, F.J. Low and Harold Johnson established a way of detecting them by the visible light that was redshifted into the infrared. A statistically valid sample of 'high redshift' quasars traced their overall distribution. There were none in our immediate neighbourhood, and only a few within several billion lightyears. Using the measure of redshift as a factor (designated 'z') rather than a percentage, 3C273, with $z = 0.16$, was the nearest at about 2 billion lightyears. The population progressively increased 1,000-fold in the $1.0 < z < 2.5$ range, and then rapidly declined.* To Schmidt, this was evidence of an evolutionary trend: "They evolved, died, and now are very rare. Probably they evolved into something else. I think they probably are the nuclei of galaxies, but mind you, that's speculation."

In 1943 C.K. Seyfert published a list of 12 otherwise normal spirals that had unusually bright nuclei. The most prominent case cited was M77 in Cetus. Others were found later, some of which were radio sources. One, 3C120 in Camelopardalis, bore a striking resemblance to a quasar, but with only a 3 per cent redshift, leading to speculation that Seyferts might be the 'missing link' between quasars and normal galaxies. Such a relationship was strengthened by the fact that the luminosity of their nuclei can vary on a time-scale of a few weeks. If one were to be viewed from afar then, it was argued, only its nucleus would be visible and it would be classified as a quasar. Seyferts were later classified as a subtype of a broader group of galaxies with 'active nuclei', most of which were ellipticals with jets. As an alternative to the evolutionary theory, Seyferts might actually *be* quasars whose bright accretion disks and jets are partially masked.

SEEING RED

There was some early criticism. In 1965 James Terrell of the Los Alamos National Laboratory in Arizona denied that quasars were at cosmological distances. He asserted they were stars that had been ejected at high speed from the core of *our own* galaxy and, by virtue of being nearby, were not outlandishly energetic – an idea that was elaborated upon by Fred Hoyle and Geoffrey Burbidge. While this explained why most quasars were receding from us, it raised the question of why only our galaxy seemed to have ejected quasars. If M31 was similarly surrounded by quasars, most of them would be *blueshifted* from our perspective. In 1961 C.R. Lynds at the National Radio Astronomy Observatory at Green Bank in West Virginia refined the position of 3C231 in Ursa Major. The 'box' was dominated by the spiral M81, but the source proved to be M82 – an irregular member of the same cluster. In the fall of 1963 Sandage took a deep plate through the 200-inch utilising a filter to emphasise hydrogen emission rather than starlight. In addition to 1,000-lightyear-long streams of hydrogen (corresponding to the radio lobes) projecting far above and below its nucleus, distorted lanes of dust and gas were silhouetted against it, giving the impression that its nucleus was undergoing a violent explosion. Whereas Sandage sought uniformity in galaxies in order to employ them as standard candles, H.C. Arp at Caltech zeroed in on those that appeared strange. In 1966 Arp published an *Atlas of Peculiar Galaxies* that was drawn mostly from his perusal of the plate archives at

---

*   For a decade the record was $z = 3.5$, then one with $z = 4.4$ was spotted in 1987; the current record is about $z = 5.5$.

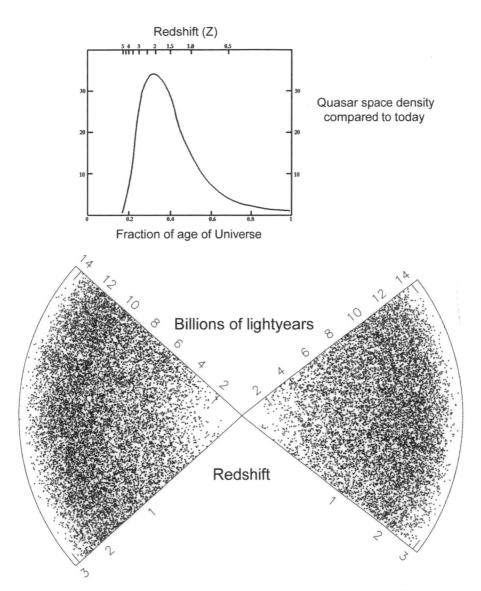

Surveys show an increase in the quasar population with increasing redshifts until the 'redshift cut off' at about $z = 2.5$ corresponding to a time when the Universe was only about one-third of its current age. (The transformation from redshift to time assumes a low-density Universe.) This is illustrated by the graph based on a survey by Peter Shaver of the European Southern Observatory. Quasars were much more common in the early Universe. The 'bow-tie' is the result of the 2-Degree Field Quasar Redshift Survey with the Anglo-Australian Telescope, completed in April 2002 by a team led by Brian Boyle in Australia and Tom Shanks of Durham University in England. The distribution of the 22,000 quasars in the two 75-by-5-degree arcs of sky ranging out to $z = 3$ clearly illustrates that quasars 'switched off' in recent times.

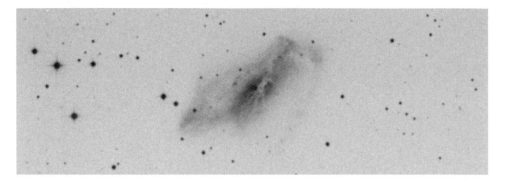

The optical counterpart of the radio source 3C120 as documented by the Palomar Observatory Sky Survey. Notice its extremely bright (star-like) nucleus (shown dark in this negative image) and the irregular wispy peripheral structure.

Mount Wilson and Mount Palomar, but supplemented by some plates that he had taken himself on the 200-inch. A few months later, Arp received a letter from J.L. Sérsic in Argentina drawing his attention to the fact that several of the objects in the *Atlas* were very close in the sky to discrete radio source. On a frustratingly thundery night on Mount Palomar, Arp went to the library and soon identified a dozen peculiar galaxies that were almost coincident with quasars. Could they *all* be chance alignments? If a quasar was *physically* related to a galaxy, any 'excess' in its redshift must be spurious. Arp drew attention to what appeared to be halos of faint stars, gas and dust far beyond the apparent outlines of peculiar galaxies, and streamers of gas that seemed to either 'point' towards – or in some cases form a 'bridge' to – a quasar. With the 'disturbed' form of M82 as evidence that galactic nuclei could undergo explosions, Arp launched a vigorous campaign arguing that quasars were objects that had been ejected from the nuclei of their host galaxies.

One of the leading critics of Arp's argument was J.N. Bahcall, a physicist at the Institute for Advanced Study at Princeton. Despite Arp's citation of a few dozen cases in which a quasar seemed to be associated with a galaxy, if the quasar redshifts really were cosmological then it would be very difficult to find one residing in a cluster of galaxies. The closest quasar is at a distance of 2 billion lightyears, and the population does not increase until about 5 billion lightyears, at which distance a normal galaxy is too faint to be readily seen. Nevertheless, Bahcall and his wife Neta, also at Princeton, found an example of a fairly nearby quasar situated in a cluster of galaxies having the same redshift, indicating that *they* were physically associated. The American Association for the Advancement of Science in Washington, DC, arranged a debate between Arp and Bahcall on 30 December 1972. Arp showed many photographs of 'pointers' and 'bridges' suggestive of associations with a variety of objects that showed different redshifts, arguing that they could not *all* be chance alignments. Bahcall argued that they *were* all chance alignments by pointing out that the sky was so dense with faint extragalactic objects that a statistically significant case could be made for almost any kind of association. Bahcall then pointed to the sheer variety of types of linkage on Arp's list as a weakness, because it

In 1963 A.R. Sandage took a deep plate of the irregular galaxy M82 in Ursa Major on the 200-inch and found a network of dark dust lanes silhouetted against its bright nucleus. This picture was taken with the 0.9-metre telescope of the Kitt Peak National Observatory. The overlay by the Hubble Space Telescope shows the dust lanes in greater detail and resolves in excess of 100 young bright compact star clusters each with 100,000 stars (dubbed 'super star clusters') in the central region. Courtesy of N.A. Sharp, AURA/NOAO/NSF and R. de Grijs of the Institute of Astronomy at the University of Cambridge, ESA and STScI/NASA.

When the National Radio Astronomy Observatory mapped streamers of hydrogen in the cluster containing M81 and M82, a connecting network was revealed which also linked to NGC3077. The radio map (lower right) is depicted here in comparison to the Palomar Observatory Sky Survey negative (lower left). Such streamers indicate a recent interaction. This is undoubtedly the reason for M82's intriguing character. The contextual image showing M81 (right) and M82 (left) is courtesy of Robert Gendler (the copyright owner) with thanks.

required a multitude of *physical* mechanisms. In any case, support for the idea of explosions waned when Philip Morrison at MIT established in 1976 that M82 was *not* undergoing an explosion. When a subsequent radio study revealed streamers of hydrogen between M81 and M82, it became evident that a recent encounter between them had prompted an intense burst of star formation in M82, and the term 'starburst galaxies' was coined. The demise of the idea of exploding nuclei undermined Arp's argument that quasars were ejected from galaxies. Despite lingering criticism by Arp's supporters, it is generally accepted that quasars *are* at cosmological distances.

By the mid-1980s it seemed likely that a quasar simply marked an early stage in the history of a galaxy in whose core there happened to have formed a supermassive black hole. The tremendous energy output was sustained for a few billion years, until the black hole had consumed everything within its gravitational reach and it had switched off. It was speculated that active galaxies probably had quiescent quasars lurking in their cores.

SLOWING DOWN

Given the value of Hubble's constant (denoted H_0) it is possible to extrapolate back to a time when all the galaxies were together, and thereby calculate the *age* of the Universe. However, this presumes that the expansion rate has been constant, which *cannot* have been so because the mutual gravitational attraction of the mass of the galaxies must act to slow it. The rate of slow down is measured by the 'deceleration parameter' (denoted q_0) and its value determines the fate of the Universe. If it were negative, some force would have to be accelerating the expansion, but it seemed safe to reject this. If $q_0 = 0$, there is so little mass-energy that gravity has had negligible effect and the age of the Universe corresponds to the 'Hubble Time' derived by extrapolating H_0 backwards. For $q_0 > 0$, this calculation establishes the *maximum* age, because the rate of expansion must have been greater in the past than it is today, and the Universe would not have taken that length of time to reach its current size. The value of q_0 determines whether the Universe is 'open', 'closed' or 'flat' in the Friedmann sense. If $q_0 < 0.5$ there is insufficient mass-energy for gravity to halt the expansion and the 'open' Universe will expand indefinitely. If $q_0 > 0.5$, there is so *much* mass-energy that gravity will not only halt the expansion but also reverse it and 'close' the Universe in a 'Big Crunch'. If, however, $q_0 = 0.5$, the Universe is 'flat' and while the expansion will be halted, doing so will take for ever, leaving no time for a contraction phase.

Any variation in the rate of expansion would show up as a difference in the value of Hubble's constant over time. Hubble had sought a departure from linearity at the far end of the redshift–distance relationship but there was considerable scatter in his data, and in any case his survey did not extend deep enough into space. In 1972 (after extending the relationship to a recessional velocity of 250,000 kilometres per second) Sandage perceived a hint that the Universe might be 'closed', but as the 'signal' was still almost indistinguishable from the 'noise' he warned that this conclusion was subject to review. In light of this difficulty, he decided to address the problem differently. As the distances to the nearest clusters were known with greater accuracy, he sought evidence of departure from 'Hubble flow' nearer home. The gravitational attraction between the members of a cluster of galaxies tends to resist the tendency for them to move apart as space expands. The motions of the members of a cluster therefore do not accurately reflect the Hubble constant, which is best measured by how *clusters* move apart. Measuring the nearby clusters *very accurately*

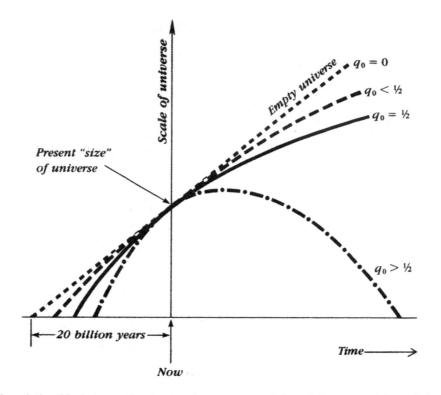

The relationship between the deceleration parameter (q_0) and the age and fate of the Universe. A value of $q_0 = 0$ corresponds to no deceleration. The Universe is 'open' if $q_0 < 0.5$ and is 'closed' if $q_0 > 0.5$; the critical value $q_0 = 0.5$ implies a 'flat' Universe.

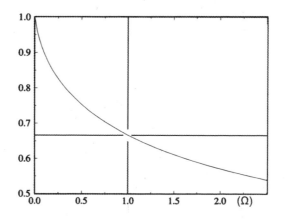

A plot of the actual age of the Universe after deceleration expressed as a fraction of the 'Hubble Time'. For a 'flat' Universe with $\Omega = 1$, the Big Bang occurred two-thirds of the age with no acceleration.

might reveal departure from linearity. The magnitude of the departure would be much smaller than that at the far end of the relationship, but the data would be slightly better and it might show up. Sandage and Tammann joined forces with the Chilean astronomer Eduardo Hardy and studied the apparent brightnesses of the 30,000 galaxies in Zwicky's catalogue (covering the 30 nearest clusters distributed all across the sky) and found that they all fitted *precisely* on the linear relationship. The mutual gravitation between the clusters was evidently insignificant, which contradicted the previous hint that the Universe was measurably 'closed'. However, in 1974 Jay Frogel at Ohio State University reported an infrared study showing that the giant ellipticals were composed primarily of red giants. This enabled Sandage to recalibrate his standard candles to take into account the fact that they dimmed significantly with age, and hence with redshift, and this eliminated the tentative evidence for the Universe being 'closed'.

In a paper in the *Astrophysical Journal* in 1974, J.R. Gott and J.E. Gunn at Caltech and B.M.H. Tinsley and D.N. Schramm at the University of Texas decided that instead of trying to measure departure from Hubble flow they would determine

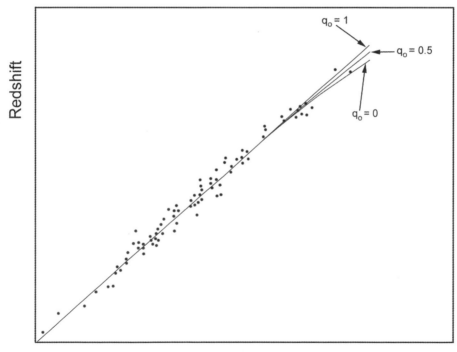

Apparent brightness

In principle, measuring the value of Hubble's constant at distances corresponding to an early epoch should reveal the value of the deceleration parameter (q_0), but in the early 1980s, having calibrated 'first ranked' giant ellipticals in clusters as 'standard candles', Sandage found that an extended plot of redshift versus apparent brightness ranging as faint as 18th magnitude did not distinguish between the alternative outcomes.

whether the mean mass-energy density of the Universe (denoted 'Ω') was sufficient to halt and reverse the expansion in a 'closed' Universe. After ruling out a variety of places in which mass might have been 'hidden' from their survey* they found that Ω was in the range 0.01 to 0.1, and concluded that the Universe was 'open'. They used three methods. First, the total luminosity of galaxies enabled the mass of the stars emitting that light to be estimated; this gave $\Omega = 0.01$. On the other hand, measuring how galaxies in clusters influence one another gave an estimate of their total masses, rather than just their luminous mass; this gave $\Omega = 0.1$. This led to the conclusion that only 10 per cent of the gravitational attraction derived from mass that radiated light, the rest was 'dark'. In 1973 the Copernicus Observatory measured absorption lines in interstellar space and established that the deuterium abundance is of the order of 1 in 50,000. As the ratio of deuterium to helium was extremely sensitive to conditions in the primordial fireball, this measured the mass-energy density in the aftermath of the Big Bang, rather than as it is today; it, too, gave $\Omega = 0.1$. If the Universe was 'flat', then 90 per cent of the gravitating mass-energy was 'missing'.

INFLATION

As the year 1979 was the hundredth anniversary of Albert Einstein's birth, a centenary volume of review papers edited by S.W. Hawking and Werner Israel was published. R.H. Dicke and P.J.E. Peebles presented a paper entitled 'The Big Bang Cosmology – Enigmas and Nostrums'. After reflecting on the success of the 'standard model' of the Big Bang, they pondered the fact that the Universe appears to be *almost* flat and *extremely* homogeneous. Friedmann's equations require the slightest deviation from $\Omega = 1$ at the moment of creation to *rapidly* diverge and become either infinitesimal or huge. They argued that if Ω was 0.1 today, then for its value to have diverged so little it would have to have been within one part in 10^{14} of 1 when the Universe was only 1 second old, which strongly implied that Ω was *precisely* 1 and the rest of the mass-energy required for the critical density remained to be found. What, they asked rhetorically, had established this critical balance? When Dicke raised this question at a seminar at Cornell in the spring, A.H. Guth, a particle physics postdoc, was very impressed, but as he was not particularly interested in cosmology he filed this fact away in the back of his mind – half forgotten. At that time, mainly as a result of the prompting of his colleague S.-H.H. Tye, Guth was trying to calculate how many 'magnetic monopoles' were created as the Higgs field 'froze' at the conclusion of the era of grand unification.

J.C. Maxwell was puzzled by the fact that no matter how finely he divided a bar magnet it merely produced a pair of smaller magnets, each of which had north–south poles. While his equations allowed for isolated poles, he reasoned that if magnetic fields were due to the motion of electric charges – that is, electric current – then there

---

* For example, it was possible that most of the mass of the Universe is present in the form of gas and dust in intergalactic space, but the fact that quasar light seems to reach us without significant modification is an indication that the space between clusters is truly empty.

would be no *need* for magnetic monopoles. In 1931 P.A.M. Dirac discovered that there *should* be a unit of magnetism, and that quantum mechanics provided the *ratio* between the units of electric and magnetic charge. Furthermore, he calculated that the attraction between two magnetic monopoles would be 60 times stronger than that of the strong force, making it the strongest of the fundamental forces. In the 1960s, it was thought that monopoles would be point-like particles with masses of a few GeV which, while heavy for a particle, was not excessive. It was suggested that if such a particle was adrift in the magnetic field that (although weak) pervades the galaxy, it would be readily accelerated to relativistic speed and so be present in cosmic rays. Upon penetrating the atmosphere, such an energetic particle would strip electrons from a succession of atoms in its path. Searches to find such spectacular ionisation trails failed. With each close encounter, the monopole would lose a little of its kinetic energy and slow down. Since air is not very dense, it was argued that if a monopole could be brought to a halt in a dense medium it might be possible to detect it directly. Perhaps they had collected in rocks that had been exposed at the surface for a long time. However, studies of rocks (some of which had been retrieved from the Moon by astronauts) proved negative. It was concluded, therefore, that if monopoles existed, they were extremely rare.

Gerard 't Hooft in Holland and Alexander Polyakov independently suggested in 1974 that magnetic monopoles would be produced during the spontaneous breaking of the grand unification symmetry. As point-like flaws in the Higgs field preserving a 'bubble' of false vacuum whose core was 10^{16} GeV, they would be fundamentally different to other elementary particles, and would contain a mass comparable to that of a bacterium within a volume comparable to a proton. Magnetic fields in space would not be able to accelerate such massive particles to relativistic speeds. As they penetrated the atmosphere, the fact that they were travelling slowly would give the atoms they encountered time to react, and so they would not leave ionisation wakes. And since such particles would not lose significant energy in encounters, they would not be slowed down on entering rock. It was not surprising that earlier searches had failed. Did monopoles really exist? Before techniques could be devised to seek such massive monopoles it was necessary to determine how numerous they were, which was what Guth and Tye were attempting to do. An early estimate had implied near parity between monopoles and protons, but this could not be the case; if it were so, then the gravitational attraction of the enormously heavy monopoles would have dominated the Universe and it would have collapsed within a few thousand years. Guth and Tye were intrigued that the absence of a profusion of monopoles implied that the SU(5) grand unification theory was flawed. Soon after reaching their astonishing conclusion, they discovered that J.P. Preskill, a student of Steven Weinberg at Harvard, had reached the same conclusion employing a different way of calculating the production of monopoles, so they invited him over for a discussion. In fact, Preskill's research was so far advanced that he already had a draft paper entitled 'Cosmological Production of Superheavy Magnetic Monopoles' in which he proposed that the grand unification theory was flawed. In contrast, Guth and Tye decided to accept the prediction of large numbers of monopoles and to ponder what could have happened to them; that is, to solve what they saw as the 'monopole problem'. Since monopoles are 'knots' left in the frozen Higgs field, Guth and Tye looked for ways of smoothing the field. They were struck by the phenomenon of 'supercooling'. If a glass of water is dynamically stable, it can be chilled below 0 °C without freezing, but as soon as it is disturbed the phase change will occur instantaneously. Might the freezing of the Higgs field have been similarly delayed, providing sufficient time for the field to smooth itself out and thereby reduce the opportunity for knots to form? In fact, Sidney Coleman at Harvard had already speculated upon a 'hang up' in the symmetry-breaking process which had let the Universe persist in its symmetric state, in a false vacuum, longer than the nominal cooling rate would permit. This prompted Guth and Tye to investigate the process more deeply, with encouraging results: the time spent in the false vacuum would indeed have enabled the Higgs field to smooth itself to reduce the number of monopoles dramatically, and those that had been created would have been scatterred far and wide. At this point (late 1979) Guth began a one-year sabbatical at Stanford University. Having reached what they considered a reasonable result, he and Tye began to write a paper. During the summing up, Tye pointed out that latent heat is liberated in a phase change. As water freezes, the molecules cease to jostle one another, settle into the crystal, and their kinetic energy is released as 'heat of crystallisation'. In supercooling, however, this energy remains latent as long as the phase change is delayed, and as long as the 'metastable' state persisted the Higgs field would have been seething with energy. What effect would this energy have had? As Tye had to make a trip to China, they put the paper 'on hold' while Guth investigated the latent heat issue.

Einstein invented the 'cosmological constant' to represent the repulsive force that he thought resisted the mutual gravitational attraction that would cause the Universe to collapse. Of course, when Hubble discovered that the Universe was expanding, Einstein dismissed the cosmological constant; the linearity of the redshift–distance relationship implied that its value was *zero*. Quantum theory, however, requires that 'empty space' – the vacuum – be seething with energy in the form of virtual particles. The Higgs field is the very essence of the vacuum. While it was in a metastable state it would still have retained the 10^{16} GeV energy density extant at the end of the era of grand unification. Working late one night in early December, Guth realised that the vacuum energy at this time would have manifested itself as a *negative pressure*. In general relativity, a pressure is equivalent to a gravitational field: a positive pressure is an attractive gravitational field and a negative pressure is a repulsive gravitational field. The Higgs field *should* have frozen at 10^{-35} second, as the temperature of the fireball fell through 10^{27} K, but while the field persisted in the metastable state, the negative pressure would have driven expansion. As a scalar field, its energy density would have been *constant* during expansion, and the *total* energy in the field would have *increased* as its volume increased. Instead of being diluted by the expansion, the total energy of the false vacuum would have dramatically increased. As a result, the radius of the Universe would have grown exponentially. Although the symmetry finally broke and the field froze at 10^{-33} second, this brief interval in the metastable state was sufficient for the radius of the Universe to *double* 100 times. Running the 'standard model' of the Big Bang backwards, the radius of the *observable Universe* was about 10 centimetres at that time. The exponential expansion during the delayed phase change could therefore have grown this from a volume no greater than that of a proton. While the subsequent expansion was more sedate (with the radius increasing in proportion to the square root of the passage of time, retarded somewhat by the deceleration parameter) this has nevertheless been sufficient to increase the radius to its current size of 10 billion lightyears. The beauty of the negative pressure was that it drove the gravitational energy of the Universe to an incredible *negative* value, and when the energised Higgs field finally collapsed, its latent energy 'condensed' into mass by pair production. In effect, these two cancel out and the total energy of the Universe is *precisely zero*! Thus, as E.P. Tryon had insightfully suggested in 1973, the Universe could have emerged from a 'vacuum fluctuation'.

At a stroke, this 'inflation' (as Guth termed it) explained away many mysteries. It *required* $\Omega = 1$, because the speck in the Universe of the grand unification era that was expanded into the observable Universe was so inflated that any curvature was flattened out – it is *effectively flat* because *any* curved surface will appear flat if it is inspected closely enough. The shape of the 'larger' Universe, beyond our 'horizon', is spherical. Despite observational evidence for $\Omega < 1$, this was simply not tenable. As Dicke had surmised, the majority of the gravitating mass-energy in the Universe was 'missing'. Furthermore, as inflation meant that the observable Universe derived from a tiny speck of the grand-unification-era space (rather than being *all* of it), the original volume would have been so small that it would not have spanned very far across any gradients in the fields. Also, since any irregularities were smoothed out during inflation, the Universe was rendered homogeneous and isotropic *prior* to the onset of Hubble flow. Einstein had postulated a homogeneous Universe for the sake of simplifying his model and at the time this had appeared to be a ridiculous oversimplification but in retrospect, in light of inflation, it was realised to be a *precise statement* of conditions in the aftermath of the inflationary episode. All of this dawned on Guth more or less overnight. Some time later he realised that inflation also resolved the 'horizon' issue by which, despite the fact that light from one side of the Universe has not yet had time to cross to the other, the laws of physics are the same everywhere. This is the case because at the onset of inflation at 10^{-35} second the observable Universe was no larger than 10^{-15} metre across, which light could have spanned. Furthermore, since inflation smoothed out the Higgs field, it probably did so *without* leaving monopoles when it finally froze. In January 1980, after six weeks of working through the implications of the theory in isolation, Guth explained it to Coleman, who promptly expressed his approval. After presenting his findings in a seminar at Stanford, Guth set off back east on a tour of academic establishments to spread the word. In general, particle physicists were appreciative. Cosmologists mulled over the implications for some time before signing up, but did so because it resolved many otherwise intractable problems. By the time Guth sent in the first of a string of papers in August, he had accepted a permanent post at MIT.

As Guth refined his inflationary theory, he realised that although the Higgs field would have been considerably smoothed while in the metastable state, there would

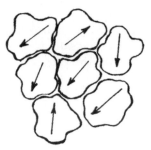

In A.H. Guth's model of inflation, the Higgs field suddenly froze after a period in a metastable state, and the simultaneous spontaneous symmetry breaking in different regions would have resulted in a number of bubbles in which the randomly selected 'direction' for the field was different, and which consequently would not have been able to coalesce seamlessly.

inevitably have been random fluctuations. When the phase change finally occurred, it would have been virtually instantaneous and would have started simultaneously in different places, producing 'bubbles' of true vacuum within which the 'arrows' of the Higgs field were differently aligned. As Guth explored this process, he realised that the fluctuations in the field would have prevented these bubbles from coalescing seamlessly.

Guth's concern was relieved in December 1981 when he happened across a paper by A.D. Linde, a theorist in Moscow, who argued that the freezing of the field had proceeded *gradually*, and instead of making lots of bubbles of true vacuum as it froze it had produced a *single* bubble. Actually, Linde had independently discovered inflation. Furthermore, he had realised that *slowly freezing any scalar field* would give rise to inflation. It was a *second-order* phase change rather than a first-order one. This meant that there was no need to invoke a supercooling process. The crucial issue was that the transition from the false vacuum into the true vacuum proceeded sufficiently slowly for the field to produce a uniform vacuum by shedding its energy by *progressively* condensing out mass-energy rather than all at once as Guth had envisaged. In the era of grand unification, the value of the Higgs field would have been fluctuating randomly. In regions where it was far from its minimum energy state, it would have been dominated by energy whose density was decreasing only very slowly, and so it would have inflated. Because it inflated *homogeneous* regions out of primordial chaos of quantum fluctuations, Linde named it 'chaotic inflation'. The observable Universe is contained inside one of these bubbles. The *largest bubble* would be derived from the region that was *furthest* from its minimum energy state. We may or may not be in *that* bubble.

In 1982, several researchers noted that Heisenberg's Uncertainty Principle meant that even as the Higgs field was being smoothed out, it *must* have suffered random fluctuations, and so there must have been *slight* variations of density in the resulting mass-energy condensation. Had these inflated quantum inhomogeneities created the 'ripples' in the cosmic background radiation, and had these subsequently served as the 'seeds' of the largest scale structures in the Universe?

This picture of a cluster of galaxies in the constellation Centaurus was taken in 1975 with the 4-metre Blanco Telescope at the Cerro Tololo Interamerican Observatory in Chile. Courtesy of NOAO/AURA/NSF.

This picture of the cluster of galaxies in Coma Berenices was taken in 1974 with the 4-metre Mayall Telescope of the Kitt Peak National Observatory. Courtesy of NOAO/AURA/NSF.

This picture of Abell 2151, a small irregular cluster of galaxies in Hercules, was taken in 1974 with the 4-metre Mayall Telescope of the Kitt Peak National Observatory. Courtesy of NOAO/AURA/NSF.

This picture of a distant cluster of galaxies in the constellation Hydra was taken in 1974 with the 4-metre Mayall Telescope of the Kitt Peak National Observatory. Courtesy of NOAO/AURA/NSF.

This picture of a small part of the Virgo cluster of thousands of galaxies that spans a wide area of sky, was taken in 1974 with the 4-metre Mayall Telescope of the Kitt Peak National Observatory. Courtesy of NOAO/AURA/NSF.

LARGE-SCALE STRUCTURES

Hubble had noted that while galaxies congregate in clusters, the clusters appeared to be randomly distributed, which implied that on the largest of scales the Universe was homogeneous and the clusters were simply being carried along as space expanded. After graduating from Vassar College in 1948, V.C. Rubin studied the kinematics of galaxies for her master's at Cornell. When presented to a session of the American Astronomical Society in Philadelphia in 1950, her paper reporting that clusters themselves show systematic motions hinting at large-scale 'superclustering' (although she did not use that term) was heavily criticised.

Hubble's analysis was based on inspecting very narrow areas of the sky. In the 1960s G.O. Abell, a Caltech graduate student, gained a much broader perspective by charting the Palomar Observatory Sky Survey. In addition to discovering many more clusters (which he listed in his own catalogue), he found that although many of these clusters were compact spherical clouds of galaxies, others were irregular, and there appeared to be filamentary strands linking them together, implying that there was a *structure* on this largest of scales. Ya.B. Zel'dovich in the Soviet Union was an early convert to the idea that the Universe had an origin. By the mid-1960s he had a large team working on the details. In his doctoral thesis, Joe Silk had examined the effect of the intense radiation on density fluctuations in the primordial fireball, and found that it would have smoothed irregularities on scales up 10^{13} solar masses. The only irregularities likely to have survived until the fireball ceased to be radiation-dominated would have been larger than this. This implied that the structures that evolved later did so from these large lumps. In 1967, Zel'dovich reasoned from this that structures formed in a *top-down* manner, and he predicted that we ought to see an intersecting pattern of clumps and filaments composed of many clusters of galaxies.

In 1969, following up on his study of conditions in the fireball, P.J.E. Peebles created a computer model in which he observed 2,000 randomly sprinkled galaxies in an expanding space clustering in a *bottom-up* manner due to their mutual gravitation. It was suggestive, but he required real statistical data. He developed the 'correlation function' to *measure* the way in which galaxies are distributed on the sky, and then analysed a catalogue of the entire northern sky which had been compiled from 1,256 plates taken by Donald Shane and Carl Wirtanen on the 36-inch Crossley reflector at Lick Observatory and published in 1954. It took five years to prepare the data for a computer and perform the analysis, but the result was surprising, and with 1 million galaxies Peebles was sure that it was statistically significant. In effect, the correlation function measured the numbers of galaxies appearing within ever larger circles on the sky. Galaxies congregate on *all* scales, with the function smoothly declining out to a separation of 50 million lightyears, at which point it fell sharply. Interpreting this in terms of his bottom-up theory, he concluded that primordial structures smaller than this characteristic size have been 'organised' by gravitation, and only structures exceeding this size still retain their primordial form. However, when the catalogue was printed with shades of grey indicating the number of galaxies per grid square of sky, a filamentary structure

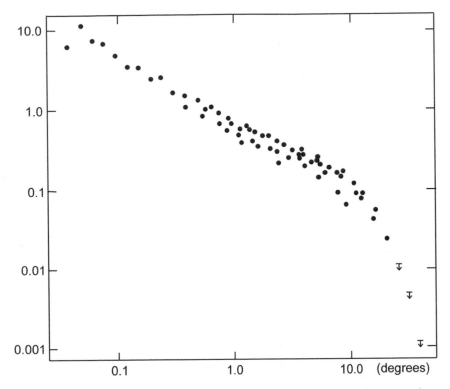

In plotting the angular correlation function (vertical axis) against the angular separation between any pair of galaxies in a comprehensive survey of the northern hemisphere, P.J.E. Peebles determined the manner in which galaxies are distributed in the sky.

could be perceived, suggestive of large-scale structures which his bottom-up theory could not explain. Was this filamentary structure real or was it merely an optical illusion? By the mid-1970s, therefore, one of the burning issues of cosmology was whether the large-scale structures were created top-down or bottom-up.

DARK MATTER

In stipulating that $\Omega = 1$, the inflationary model of the Big Bang raised the issue of the 'missing mass'. A number of strange observations had been made over the years, and their significance was belatedly appreciated.

The Coma cluster is a large conglomeration of 11,000 galaxies which spans about 8 million lightyears. Its members, being typically 300,000 lightyears apart, are more densely packed together than are the 20 members of our Local Group. In the early 1930s, soon after joining Caltech, Fritz Zwicky noted that the individual galaxies in the Coma cluster were moving at such high speeds relative to the mean that for them to remain gravitationally bound there had to be more mass than was apparent. He dubbed this 'non-luminous' matter. Although profound, this conclusion attracted little attention.

On joining the Carnegie Institution in Washington, DC, Rubin worked with W.K. Ford and built an innovative spectrograph using an image-intensifier in order to capture the spectra of faint objects. Ignoring the popular frenzy to study quasars and active galaxies, Rubin decided to investigate *normal* galaxies. Specifically, whereas Sandage integrated light from an entire galaxy in order to measure its mean redshift so as to update the redshift–distance relationship, the new sensitive spectrograph measured the variation in radial velocity across the diameter of a galaxy. The velocity distribution along the major axis, known as a 'rotation curve', is a measure of the orbital velocity of stars or gas clouds at different distances from the centre, and hence is a profile of the distribution of mass in the galaxy. In the Solar System (in which 99 per cent of the mass is in the centre, in the form of the Sun) the planets obey Kepler's laws and their orbital velocities decrease with the square root of the radius of their orbit – a profile that is known as 'Keplerian decline'. Like everyone else, Rubin presumed the distribution of luminosity in a galaxy to be correlated with the distribution of mass because, after all, the light came from stars. Rubin selected M31 as the first objective because it subtended such a wide angle that the slit of the spectrograph could isolate small sections of it to assemble the rotation curve in fine detail. The observations, made at the Kitt Peak National Observatory, came as a surprise: instead of the expected Keplerian decline, the rotation curve was 'flat': after rising rapidly, deep within the nucleus, it slowly rose to a level that persisted into the peripheral region where, even though the luminosity declined, the mass evidently did not. This result was published in 1970 but attracted little response, and for a time Rubin put the puzzle aside in order to resume her kinematical studies. After a few years, however, she began a systematic study of the rotation curves of different types of spiral and her results showed M31 to be the rule rather than the exception. In each case, the outer regions rotated as if they were contained *within* a larger structure rather than being on the periphery. By a curious coincidence, in numerically modelling the formation of galaxies, Peebles and Jeremiah Ostriker discovered in 1973 that the spiral form, which was really nothing more than a density wave, was readily disrupted unless its disk resided within a larger spherical halo. The fact that this halo was *invisible* reinforced Zwicky's insight that galaxies were mostly non-luminous.

The rotation curves of ellipticals were more difficult to measure, but the results implied that they were mostly dark matter. Dwarf galaxies containing typically only 100 million stars seemed to be full of it. Further studies suggested that rather than each galaxy having its own halo, dark matter pervaded an entire cluster and the galaxies were merely dense patches that were luminous. Astronomers had presumed that the Universe comprised what they could *see*, little suspecting that the luminous material was insignificant. What was the 'dark matter'? A variety of candidates were offered, including red dwarfs, brown dwarfs, black holes, intergalactic hydrogen and a variety of exotic particles – some familiar, some hypothetical – but the total mass-energy was still no more than 30 per cent of that for $\Omega = 1$.

An analysis of the rotation curves of spiral galaxies by V.C. Rubin revealed several contributing components, including a 'halo' surrounding the visible part of a galaxy. A 'deep' image of the M100 by David Malin of the Anglo Australian Observatory (top left) revealed that it was much larger than the familiar spiral (which has been superimposed in black) implied. A processed image of M94 taken with the 0.9-metre telescope of the Kitt Peak National Observatory by Hillary Mathis recorded such 'outer' structure in some detail. The M94 image is courtesy of N.A. Sharp, NOAO/AURA/NSF.

10

Supermassive black holes

GALACTIC CORES

As the total mass of a galaxy must far exceed that of any central supermassive black hole, the presence of such an object will make an insignificant contribution to the gravitational field that controls the motion of the material in the outer regions but it will dominate the nuclear region. In the absence of a supermassive black hole, the orbits of the stars in the nucleus will be random (in contrast to the sedate procession of stars rotating in the plane of a spiral) with the stars swarming about like bees. If a black hole is present, however, it will draw material in, and the angular momentum of that material will cause it to form a rapidly rotating disk. A disk of either gas or stars at the dynamical centre of a galaxy is therefore a tell-tale sign of a supermassive black hole's presence, as the disk would disperse without the intense gravity of the invisible monster to restrain it.

In principle, the rotation profile of the material in a galaxy's nuclear region can be measured by orienting the slit of a spectrograph across it so as to determine the distribution of radial velocities, but few ground-based telescopes have the resolution required to isolate even the nuclear region of M31, which subtends 1 second of arc. Nevertheless, in 1983 Alan Dressler at Caltech decided to look at M77 (also listed as NGC1068 and as the radio source Cetus A), which Seyfert had cited as the classic example of an 'active' galaxy and which was a strong candidate for having a central black hole. For comparison, he also inspected M31, which he had no reason to think had one. Although M77 proved too far away for the 200-inch to resolve its nuclear region, Dressler was astonished to discover a disk of stars deep in M31 circulating at 150 kilometres per second, implying a black hole of 30 million solar masses. The presence of a central black hole in a galaxy that showed no sign of *ever* having been 'active' was a major revelation. A study by John Kormendy of the University of Texas at Austin on the Canada–France–Hawaii Telescope on Mauna Kea in Hawaii not only verified M31 but also detected several others, including one of 3 million solar masses in M31's diminutive satellite M32, and one of 1 *billion* solar masses in the giant spiral NGC3115. Dressler and Kormendy were members of a team known as the Nukers,* who set out in 1994 to use the resolving power of the Hubble Space Telescope to study more distant galaxies – with immediate result. The active galaxy NGC4261 is the optical counterpart of 3C270. Team members Laura Ferrarese and Holland Ford of the Johns Hopkins University in Baltimore and Walter Jaffe of Leiden University discovered a torus of gas and dust in its core, within which VLBI resolved the accretion disk of a black hole of 500 million solar masses.

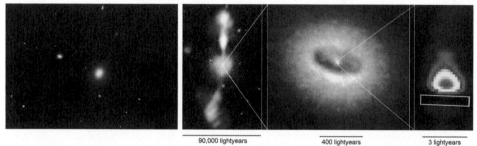

| 90,000 lightyears | 400 lightyears | 3 lightyears |

The giant elliptical NGC4261 is one of the brightest galaxies in the Virgo cluster (left). Its radio jets are depicted (centre-left) superimposed on the optical image. The Hubble Space Telescope found a torus of gas and dust in its core (centre-right). The inner structure was resolved by high-resolution radio observations during the VLBI Space Observatory Program. The outlined section is believed to appear dark because the light source in the centre of the torus is hidden by the opaque accretion disk that surrounds a black hole of 500 million solar masses. Courtesy of Laura Ferrarese and Holland Ford of the Johns Hopkins University in Baltimore, Maryland, and Walter Jaffe of Leiden University.

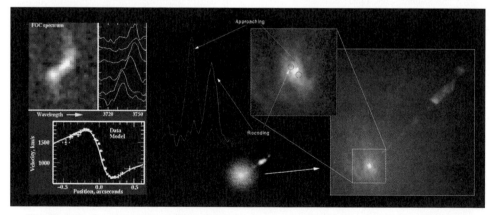

The Hubble Space Telescope found a disk of gas some 500 lightyears in diameter at the heart of the giant elliptical M87, with a radial velocity profile which implied the presence of a central black hole with the incredible 3 billion solar masses.

---

*   The name 'Nuker' was coined by Sandra Faber of the Lick Observatory when constituting the team in 1990. Given that objects in the nuclei of galaxies were the focus of the research project, she began a general letter 'Dear Nukers', which seemed natural, and the name stuck.

The radio source Cetus A lies at the heart of the bright nucleus of the Seyfert galaxy M77 (NGC1068). The Hubble Space Telescope resolved the nuclear region (insert, right). A polarisation study suggested the presence of an obscuring ring or torus of material, and this was confirmed by VLBI observations (insert left) of the radio jets. The main picture is courtesy of Michael Chase, Adam Block and NOAO/AURA/NSF.

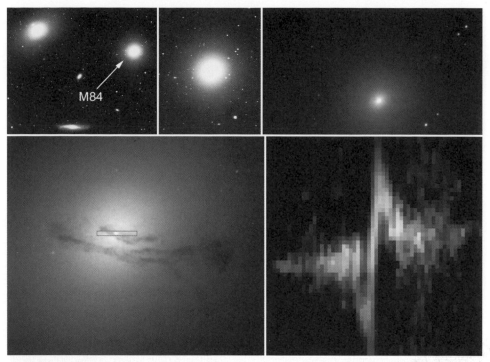

The Hubble Space Telescope's Imaging Spectrograph inspected the core of M84 in the Virgo cluster. The zig-zag in the velocity line indicated a rapidly swirling disk of material encircling an object so compact that it could only be a black hole of about 300 million solar masses. The CCD shot of the nucleus (top right) is courtesy of NOAO/AURA/ NSF. The contextual views (top left and centre) were taken with the 4-metre Mayall Telescope of the Kitt Peak National Observatory.

The pace of supermassive black hole discoveries picked up with the installation of the Space Telescope Imaging Spectrograph in early 1997, and the Nukers eagerly sought a correlation with the host galaxy in order to gain some insight into how the central presence related to its host. The various morphologies of galaxies derive from dynamical considerations. A galaxy such as our own, or M31, comprises an almost spherical nuclear region (dubbed the 'bulge') whose centre is coincident with that of a broad disk of clouds of gas and dust and the stars created from them. An elliptical galaxy has no disk, it is 'all bulge'. This is not to say that ellipticals are smaller than spirals, as giant ellipticals like M87 are *larger* than most spirals. Some galaxies, such as M33, have such insignificant bulges that they are referred to as 'disk galaxies'. The Nuker team soon noticed that they were finding central black holes in ellipticals and spirals with bulges, but not in disk galaxies. It was also apparent that the larger galaxies tended to have more massive black holes. In 1993, Kormendy noticed that there was a linear relationship between the mass of the black hole and the luminosity (and hence the mass) of a galaxy's bulge – with the black hole contributing about 0.2 per cent of its mass. Fellow Nuker, Karl Gebhardt of the University of California at

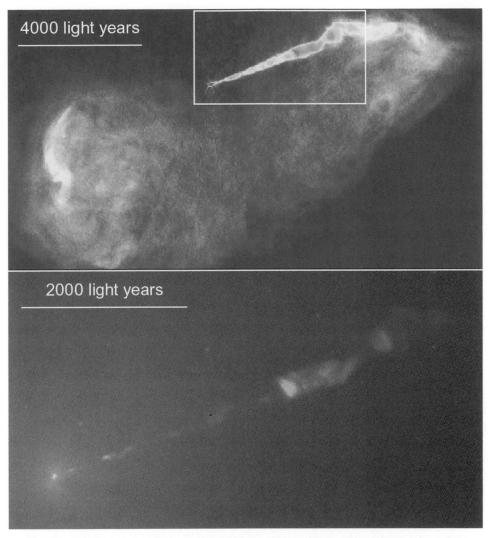

4000 light years

2000 light years

The Very Large Array radio image (top) shows the prominent jet and radio lobes of the giant elliptical M87 in Virgo. The Hubble Space Telescope provided an optical view of the nucleus and the structure within the jet.

Santa Cruz, then realised that the mass of the black hole was related to the average speed of the stars in the *main body* of the host – with the most massive black holes being found in galaxies in which the stars were travelling most rapidly. Since it had been believed that the material beyond the nucleus would not 'feel' the presence of the black hole, this discovery came as a surprise to the team.

In fact, this relationship had been predicted several years earlier by Joseph Silk and Martin Rees. It had initially been presumed that the primordial gas had simply condensed to form stars and galaxies, and the debate focused on whether this had

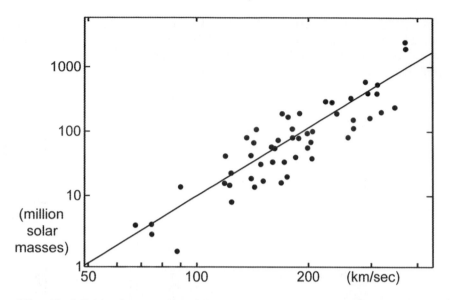

When Karl Gebhardt, a member of the 'Nuker' team, plotted the inferred masses of the black holes in the cores of galaxies against the average speed of the stars located in the outer part of the galaxies – where they should not be able to directly sense the mass of the central black hole – he was surprised to find a linear relationship.

occurred in a top-down or a bottom-up manner. Silk, however, had proposed that the centre of each primordial gas cloud collapsed to create a black hole which, as it fed on the nearby gas, 'switched on' as a quasar, and the radiation pressure induced shock waves that simulated star formation. The process of galaxy formation was therefore *a result of* the black hole's presence. Its quasar phase ended when the black hole ceased feeding, which the theory predicted would happen when the quasar grew so bright that its radiation literally drove the rest of the material beyond the black hole's reach. Silk found that this process depended on how rapidly the stars in the outer part of the galaxy were moving: the faster these stars were circling, the more difficult it would be to push them away, and the more powerful the quasar would need to be to produce the radiation to overcome their angular momentum. Lacking statistics of supermassive black holes in the cores of galaxies, Silk had had no way to test this prediction, but it was precisely what Gebhardt had found. This discovery was announced at the Rochester meeting of the American Astronomical Society in June 2000, and later published in *Science*. It caused a sensation as it stimulated a new view of galaxy formation. Within the vast clouds of gas that collected around the slight irregularities in density left by the primordial fireball, localised concentrations collapsed to form black holes whose gravitational attraction drew in the nearby gas to form a swirling cloud. As the black hole fed, the radiation pressure of the quasar process triggered spectacular bursts of star formation, sparking the galaxy into life. As the black hole grew, the intensity of the quasar increased. When the radiation pressure drove off the rest of the galaxy, the black hole was starved and the quasar was extinguished. Given Sandage's inference that the 'first ranked' galaxies in a cluster were of the same luminosity, this implied that the process of galaxy formation imposed a *limiting* mass. But how large were the black holes that seeded galaxies?

Astronomers had sought evidence of black holes in globular clusters in the 1970s, but ground-based telescopes did not have the resolution to study the motions of stars in their cores. This was, however, possible by use of the Hubble Space Telescope. Mayall II, also listed as G1, is a globular cluster 170,000 lightyears from the centre of M31 (approximately the same distance the LMC is from our own galaxy). At 10 million solar masses, it is the largest known such cluster, and is almost a very small irregular galaxy. An intermediate black hole of 20,000 solar masses was found in its core by a team led by Michael Rich of Columbia University. Globular clusters containing well-fed black holes may well have served as the seeds for galaxies. Observations by the Röntgen satellite in the early 1990s identified ultra-bright X-ray sources in some starburst galaxies, and a decade later the Chandra X-Ray Observatory established that these were intermediate-mass black holes.

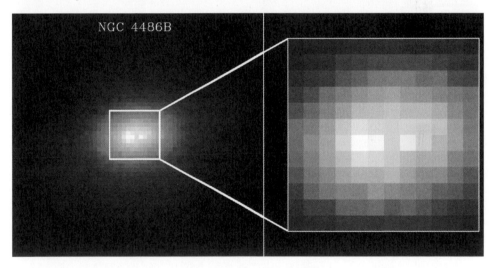

The Hubble Space Telescope discovered that NGC4486B, the satellite of M87, has a double nucleus. Courtesy of Tod Lauer of the National Optical Astronomy Observatories.

The fact that a black hole starved when its quasar drove the supply of material away begged the question of what happened when galaxies collided and merged. In 1997 Karl Gebhardt and Tod Lauer reported that NGC4486B (a dwarf companion of M87) has a pair of nuclei 40 lightyears apart orbiting a common centre, and must be the result of a merger. In 1998 Genbardt reported a Keck Telescope investigation of the dynamics of M87's globular clusters which distinguished two families. Despite their symmetry, ellipticals would appear to be mergers.

After a radio survey in the late 1960s, J.G. Bolton in Australia drew attention to NGC6240 in Ophiuchus. At Bolton's suggestion, R.A.E. Fosbury examined it using

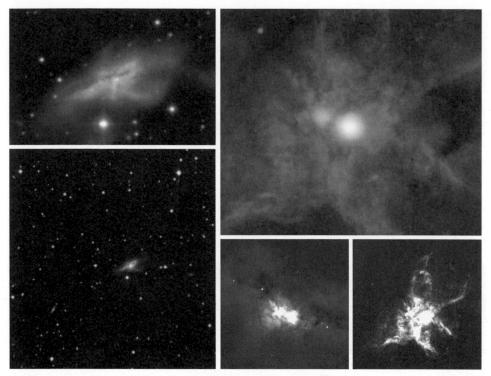

The radio galaxy NGC6240 in Ophiuchus was discovered to have a strong infrared excess. A close inspection revealed it to possess a double nucleus. After kinematic studies produced confusing results, the high-resolution imaging capability of the Chandra X-Ray Observatory found the nucleus to comprise a pair of supermassive black holes in the range 10 to 100 million solar masses, orbiting each other some 3,000 lightyears apart. This discovery supported the proposal that after quasar activity ceases, the starved central black hole continues to grow by coalescing with its counterparts in other galaxies during collisions and mergers. The context (lower left) is taken from the Palomar Digital Sky Survey. The optical image (upper left) was taken with the 2.2-metre telescope at La Silla in Chile and is courtesy of W.C. Keel of the University of Alabama and ESO/MPI. The two Hubble Space Telescope images of the central region (lower right) are courtesy of R.P. van der Marel and J. Gerssen of NASA/STScI and the Chandra overlay (upper right), depicting the X-ray emitting black holes, is courtesy of Stefanie Komossa of the Max Planck Institut für extraterrestrische Physik at Garching in Germany and NASA/CXC.

the 3.6-metre telescope at the European Southern Observatory in 1975 and found a disturbed morphology. After integrating the radio and emission-line spectrum with J.V. Wall at the Mullard Radio Astronomy Observatory in Cambridge, he suggested that it was a collision between gas-rich galaxies. In 1978 the IRAS satellite found it to be extremely luminous in the infrared. A kinematic study reported in 1991 by J. Bland–Hawthorn at Rice University in Houston, A.S. Wilson at the University of Maryland and R.B. Tully at the University of Hawaii, found it to comprise a pair of disk galaxies that began to merge about 30 million years ago. This was supported by infrared observations several years later by a team led by P.P. Van der Werf of the Max Planck Institut für extraterrestrische Physik at Garching in Germany. Stars are forming, evolving and exploding at an exceptional rate. Radio, infrared and optical all indicated a pair of bright nuclei. The Röntgen satellite saw two X-ray sources in the central region, but the situation remained enigmatic. Bland–Hawthorn had tentatively suggested that the core contained a black hole of *at least* 40 billion solar masses, but a later radial velocity study by the Hubble Space Telescope Imaging Spectrograph did not support such a massive central feature. However, in 2002 Stefanie Komossa, also in Garching, led a team that used the high-resolution imaging capability of the Chandra X-Ray Observatory to find out which, if either, nucleus was an active supermassive black hole and, to their astonishment, they found that *both* were. They are orbiting one another some 3,000 lightyears apart. This explained both the confused motions observed by the Hubble Space Telescope and the very high initial estimate of the central mass. In fact, each black hole is 10 to 100 million solar masses. They are slowly spiralling towards each other, and when they meet and coalesce in a few hundred million years the fireworks display should be spectacular! This discovery supported the proposal that after quasar activity ceases, the starved central black hole continues to grow by coalescing with its counterparts in other galaxies during collisions and mergers.

Table 10.1 Supermassive black holes (circa 2001)

Galaxy		Location	Mass	Class	Comment
Milky Way		–	3.7×10^6	Sbc	
NGC205	M110	Andromeda	$< 1 \times 10^5$	Sph	Satellite of M31
NGC224	M31	Andromeda	3.5×10^7	Sb	
NGC221	M32	Andromeda	3.4×10^6	E2	Satellite of M31
NGC598	M33	Triangulum	$< 1 \times 10^3$	Scd	
NGC821		Aries	3.9×10^7	E4	
NGC1023		Perseus	4.4×10^7	S0	
NGC1068	M77	Cetus	1.4×10^7	Sb	Seyfert galaxy
NGC2778		Lynx	1.8×10^7	E2	
NGC2787		Ursa Major	4.1×10^7	SB0	
NGC3031	M81	Ursa Major	6.8×10^7	Sb	
NGC3115		Sextans	1.2×10^9	S0	
NGC3245		Leo Minor	2.1×10^8	S0	
NGC3377		Leo	1.1×10^8	E5	
NGC3379	M105	Leo	8×10^7	E1	
NGC3384		Leo	1.4×10^7	S0	
NGC3608		Leo	1.1×10^8	E2	
NGC4151		Canes Venatici	–	–	Seyfert galaxy
NGC4258	M106	Canes Venatici	4.2×10^7	Sbcp	Peculiar
NGC4261		Virgo	5.0×10^8	E2	
NGC4291		Draco	2.0×10^8	E2	
NGC4342	IC3256	Virgo	3.0×10^8	S0	

Galaxy	Location	Black hole	Mass	Class	Comment
NGC4374	M84	Virgo	8×10^8	E1	
NGC4395		Canes Venatici	$< 8 \times 10^4$	Sm	
NGC4438		Virgo	–	P	Active galaxy
NGC4459		Coma Berenices	7.0×10^7	SA0	
NGC4473		Coma Berenices	8×10^7	E5	
NGC4486	M87	Virgo	3.3×10^9	E0	Virgo A
NGC4486b		Virgo	5.8×10^8	dE1	Satellite of M87
NGC4564		Virgo	5.7×10^7	E3	
NGC4594	M104	Virgo	1.0×10^9	Sa	The Sombrero
NGC4596		Virgo	8.0×10^7	SB0	
NGC4649	M60	Virgo	2.0×10^9	E1	
NGC4697		Virgo	1.7×10^8	E4	
NGC4742		Virgo	1.4×10^7	E4	
NGC4945		Centaurus	1.4×10^6	Scd	
NGC5128		Centaurus	2×10^8	S0	Centaurus A
NGC5845		Virgo	2.9×10^8	E	
NGC6251		Ursa Minor	6.0×10^8	E2	
NGC7052		Vulpecula	3.2×10^8	E4	
NGC7457		Pegasus	3.6×10^6	S0	
IC342		Camelopardalis	$< 5 \times 10^5$	Scd	
IC1459		Grus	2.0×10^8	E3	
NGC7078	M15	Pegasus	4×10^3	–	Globular
Mayall II	G1	Andromeda	2×10^4	–	M31 Globular

Note: Black holes are expressed in terms of solar masses.

If the relationship between the mass of a galaxy and its central black hole could be extended to quasars, then the population statistics dating back to the high redshift 'cut off', so painstakingly compiled by Maarten Schmidt, would show how a feeding black hole affected its host during its quasar phase. What was required was a nearby quasar that could be studied in detail.

Cygnus A was first noted as an X-ray source by the Uhuru satellite. Data from the ANS satellite implied that the X-ray source was diffuse. This was confirmed by HEAO-1, which established that the emission was a thermal plasma co-located with the radio source, but rather larger. The imaging system on the Einstein Observatory found X-ray emission from the cluster of galaxies within which Cygnus A resides, but not from an active nucleus coincident with the radio source. In fact, there was no sign of emission specifically from the site of the radio source. Nevertheless, in 1994 K.A. Arnaud of the University of Maryland announced that the ASCA satellite had detected the presence of an X-ray point-source coincident with the radio source, and that the situation was consistent with an active galactic nucleus observed through an opaque torus of gas. This was supported by the Hubble Space Telescope securing an ultraviolet spectrum which closely resembled that of a quasar – the ultraviolet being from the 'cooler' region of the accretion disk. This conclusion was reported in 1994

When the Hubble Space Telescope inspected NGC7052 (an elliptical in Vulpecula), it discovered a black hole of 300 million solar masses encircled by a dusty disk some 3,700 lightyears in diameter. Although NGC7052 is a strong radio source and has two oppositely directed jets emanating from the nucleus, the fact that these are *not* perpendicular to the disk may indicate that the black hole and its torus do not share a common origin. One possibility is that the dust derives from a recent collision in which NGC7052 consumed a small neighbouring galaxy; if so, it will be consumed in several billion years. Courtesy of NASA, R.P. van der Marel (STScI) and F.C. van den Bosch (University of Washington).

by Anne Kinney of the Space Telescope Science Institute, who worked with Todd Hurt and Robert Antonucci of the University of California at Santa Barbara. The broad 21-centimetre absorption reported in 1995 by J.E. Conway of the Onsala Space Observatory, the Swedish National Facility for Radio Astronomy, and P.R. Blanco of the University of California in San Diego, further strengthened the case for an opaque torus.

Whereas Baade's photograph of Cygnus A had suggested the double nucleus of a collision, later astronomers had suspected an analogy with Centaurus A, in the form of a poorly resolved galaxy marred by dust lanes. Even the Hubble Space Telescope could barely resolve it but the result was intriguing: although there are dust lanes, in some regions much of the light is not from stars but from gas ionised by the nucleus. A dusty ring of young stars forms the equator of a double ionisation cone co-axial with the radio jets. In fact, Cygnus A shows all the characteristics of a quasar. This was confirmed in 2000, when the Chandra X-Ray Observatory detected hard X-rays from an unresolved point coincident with the radio source. Although they could barely resolve it, astronomers were delighted to find that there was a simmering quasar less than a billion lightyears away.

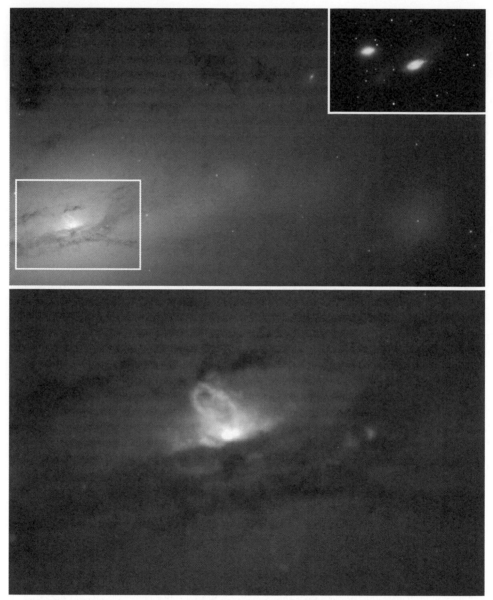

This Hubble Space Telescope image (bottom) shows the disruptive influence of the supermassive black hole in the centre of NGC4438, an active galaxy in Virgo. The prominent white patch is material swirling in the accretion disk. Twin jets sweep in opposite directions and where they impinge upon the dense slow-moving gas they generate shock waves that cause that gas to glow. In this oblique perspective, the effect of the nearside jet is readily apparent, but that on the far side is much fainter. The bright bubble is some 800 lightyears in diameter. The main imagery is courtesy of Jeffrey Kenney and Elizabeth Yale (Yale University) and NASA/STScI, and the context is by Steve and Nancy Kosar and Adam Block, AURA/NOAO/NSF.

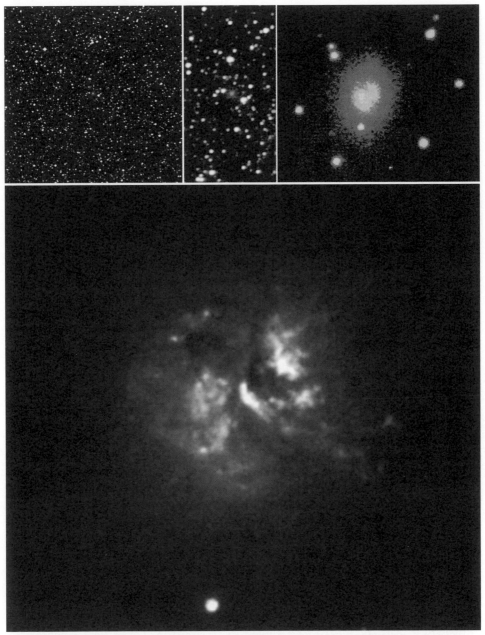

W.H.W. Baade's plate (upper left) and successive enlargements of Cygnus A, and a Hubble Space Telescope image showing the 'double nucleus' to be an illusion created by a chaotic arrangement of dust and bright patches (some of which are blue and are probably star-forming regions) all embedded in an elliptical galaxy. Courtesy of W.C. Keel of the University of Alabama and R.A.E. Fosbury of the Space Telescope European Coordinating Facility, European Southern Observatory.

Jet

The Wide-Field/Planetary Camera of the Hubble Space Telescope had been dazzled by the nearby quasar 3C273 (left). In 2003 the Telescope's new Advanced Camera for Surveys (ACS) used a coronagraph to block the glare and found the host galaxy to be more complex than expected, with a spiral plume wound around the quasar and a red dust lane. There is also a clump of material and a 'blue arc' in the path of the prominent jet. The WFPC2 image is courtesy of J.N Bahcall of the Institute of Advanced Study at Princeton and NASA/STScI. The ACS image is courtesy of André Martel of Johns Hopkins University in Baltimore and the ACS Science Team at NASA/STScI/ESA.

OUR OWN MONSTER

What about our own galaxy? The centre of the Milky Way is hidden by optical extinction. The first indication of its nature was the discovery in 1931 by K.G. Jansky that it was a strong source of radio emission. In his follow-up observations a decade later, Grote Reber designated it Sagittarius A. For many years, the centre of the galaxy remained the province of the radio astronomers. In 1974 Bruce Balick and Bob Brown, utilising the National Radio Astronomy Observatory's interferometer at Green Bank in West Virginia, found a strong point-like synchrotron source in the heart of Sagittarius A whose unusual characteristics made its nature a puzzle. They named it Sagittarius A* (abbreviated to Sgr A* – and pronounced 'A-star').

In 1970 Robert Haymes of Rice University in Houston, Texas, flew a gamma-ray detector on a balloon and noted a source in the general direction of the galactic centre. It was difficult to pinpoint, because early gamma-ray detectors did not have very good angular resolution. However, the signal was noteworthy, as it was at an energy that indicated the annihilation of electrons by antielectrons. This was confirmed by later balloon flights, but a balloon could collect data only for a short period. Sustained observations by the HEAO-3 satellite in the 1980s revealed that the source varied on a time-scale implying that it could not exceed a few 'lightmonths' in size. The gamma-ray observations were consistent with Sgr A* being a black hole. Donald Lynden-Bell at Caltech dubbed it "the beast at the galactic centre". Shortly after the gamma rays were first detected, the galactic centre was found on the Uhuru satellite's all-sky survey, but it was not until the launch of the Einstein Observatory (HEAO-2) in 1978 – incorporating the first X-ray imaging system – that it became possible to localise the source within 1 minute of arc of the galactic centre. In this part of the spectrum, its variations indicated a source of approximately 3 lightyears across.

The Very Large Array in Socorro, New Mexico, first turned to the galactic centre in 1981. Exploiting its high angular resolution, in 1983 a team led by Ron Ekers saw for the first time a 'mini-spiral' of hot gas surrounding Sgr A*. That same year, K.Y. Lo and Mark Claussen made an even more detailed map of the region, and refined the position of Sgr A*. Don Backer and Dick Sramek repeatedly measured its position in the sky in order to set a limit on its mass. If it was comparable in mass to a star, it would be seen to travel rapidly in an orbit around the centre of the galaxy, but would travel more slowly if it was very massive. After 16 years of observing it to be motionless, they concluded that it was located *at* the dynamical centre. In 1984 Farhad Yusef-Zadeh, Mark Morris and Donald Chance discovered a series of parallel filaments and threads, evidently following the magnetic field at the galactic centre. Observations in the far-infrared indicated ionised gas, seemingly in a torus some 10 lightyears across. In 1998 the Very Long Baseline Array (which used 10 sites on the North American continent to synthesise an aperture thousands of kilometers in diameter, and thereby achieve a very high angular resolution) found Sgr A* to be an *elongated* feature only a few AU on its major axis. In 1968, in making pioneering observations in the near-infrared, Gerry Neugebauer at Caltech and his student Eric Becklin had discovered a cluster (IRS16) coincident with Sgr A*. An improved detector resolved this in 1986 into red giants that were packed in together at the equivalent of 1 million stars per cubic lightyear.

The most direct way to ascertain whether Sgr A* was a massive black hole was to monitor the motions of the nearby ionised gas streamers. While the radial velocities implied that 6 million solar masses were concentrated within 10 seconds of arc of the Sgr A* radio source, the case was inconclusive because this was quite a large volume and the gravitating mass could be the star cluster that was situated within a few lightyears of the centre. The only way to be certain was to measure the motions of the stars in the cluster by high-spatial-resolution infrared *astrometry*. In the early 1990s, two teams started using 'speckle' imaging to track the proper motions of the stars deep within the cluster: one, headed by Reinhard Genzel of the University of Munich, Germany, used the 3.5-metre New Technology Telescope in La Silla, Chile; the other, led by Andrea Ghez at the University of California at Los Angeles, utilised the 10-metre Keck Telescope on Mauna Kea in Hawaii. The fact that such remote stars displayed significant proper motions confirmed the presence of a black hole in the heart of the cluster. How large was it? As the observations accumulated over the years, one star was seen to trace two-thirds of a 15-year elliptical orbit, and analysis

A wide view of the Milky Way in Sagittarius. The galactic core is obscured by gas and dust. Courtesy of David Talent at the Cerro Tololo Interamerican Observatory in Chile, NOAO/AURA/NSF.

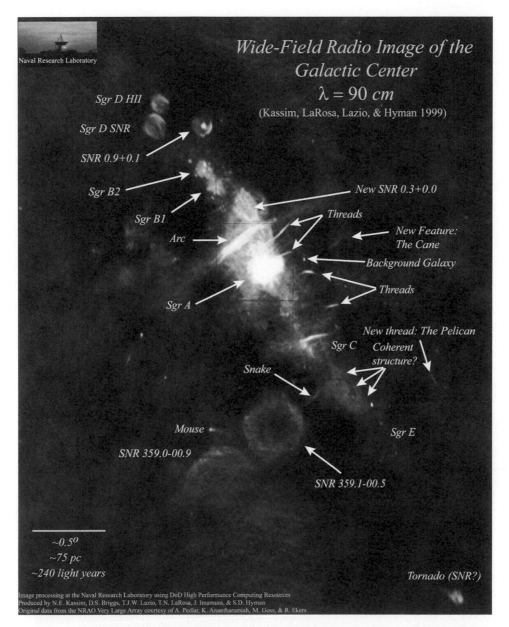

This is the largest and most sensitive radio image ever made of the central region of the Milky Way at a uniform and high resolution. It was compiled by N.E. Kassim at the Naval Research Laboratory from Very Large Array observations at a wavelength of 1 metre. The galactic plane runs from lower right to upper left and passes through Sgr A. The large bubbles are supernova remnants. The strong radio source Sgr A* (called 'Sagittarius A-star') embedded deep within Sgr A is located precisely at the dynamical centre.

of the periapsis passage of 2002 indicated a gravitating mass of 3.7×10^6 solar masses which, in such a confined space, could only be a black hole. Its event horizon must be only 15 million kilometres in diameter. The radio source seems to be ionised gas in the inner region of the accretion disk. The disk, and the elongated shape of the radio source implies that the disk is being viewed edge-on. In effect, having consumed everything within easy reach, the black hole is 'starving'. Nevertheless, an infrared source was noted at the exact site of Sgr A* from time to time during the survey, and the Chandra X-Ray Observatory has noted occasional flickering implying ongoing irregular accretion of material by the disk. Having discovered supermassive black holes in the centres of remote galaxies, astronomers now realised that they had a dormant one lurking in their own backyard, which they could study in detail right across the spectrum.

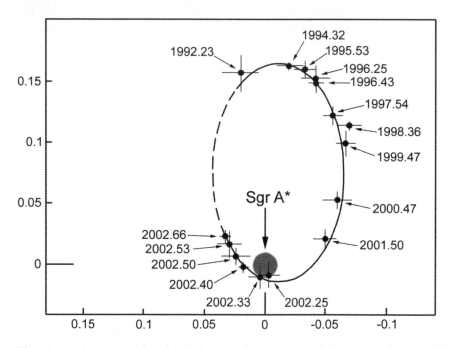

The 15-year-long eccentric orbit of the star closest to Sgr A* shows that there is a black hole of 3.7×10^6 solar masses at the dynamical centre of our galaxy. The axes are measured in seconds of arc, centred on Sgr A*. The star's periapsis passage in 2002 was at a range of 17 lighthours, or 124 AU. (Adapted from 'A Star in a 15.2-year Orbit around the Supermassive Black Hole at the Centre of the Milky Way' by R. Schödel *et al.* in *Nature*, 17 October 2002.)

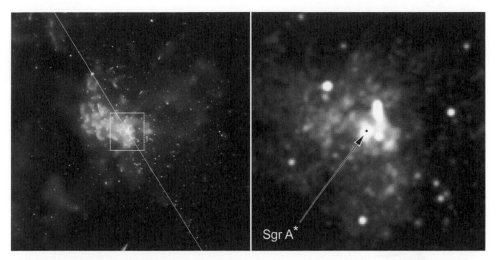

The image of the galactic centre (left) was made from the longest X-ray exposure of that region to date by the Chandra X-Ray Observatory. In excess of 2,000 X-ray sources were detected, making this one of the richest fields ever observed. The close-up (right) records emission in the 2.0–8.0 keV range from the vicinity of the supermassive black hole known as Sgr A*. Courtesy of F.K. Baganoff, NASA/CXC/MIT.

11

Fitting the pieces together

SANDAGE'S ORDEAL

J.L. Greenstein said of Sandage in 1975, "Much of what [he] is doing, he's been doing for so long that for anybody else just to catch up would take years." Even so, young astronomers questioned the very basis of Sandage's work. The disagreement burst into the open at a conference in Paris in 1976: Sandage and Tammann reported their latest results showing that Hubble's constant was 15 kilometres per second per million lightyears, but Gérard de Vaucouleurs of the University of Texas argued that it was twice this value. Whereas Sandage's value gave the maximum age of 20 billion years, the faster rate allowed only half this time, which was in conflict with a variety of other evidence. The result was a prolonged debate over which phenomena could be used as distance indicators, how they should be calibrated, and how their results should be 'weighted'. The issue was exacerbated by the fact that although Sandage's value was just 10 per cent of Hubble's original estimate, it was quoted with the same estimated error of ± 15 per cent, and there was no overlap between the ranges of the modern values. What was needed was a significant improvement in the distances of the *first steps* on the distance scale.

HIPPARCOS

By the mid-twentieth century, the distances to 10,000 stars had been estimated by measurement of parallax and various indirect methods, but any significant increase in this sample was impracticable using terrestrial telescopes.

When the European Space Agency's Hipparcos satellite was launched in 1989, it was stranded in a highly elliptical orbit by the failure of a rocket motor. However, it undertook its programme of measuring the angles between stars so that triangulation using the Earth's orbit as a baseline would enable the distances to nearby stars to be measured with an accuracy 100 times better than was previously possible. By the end

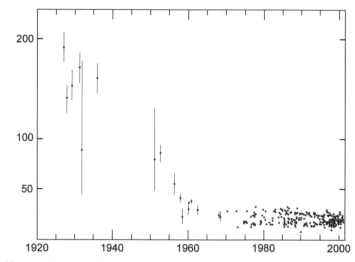

A plot of how estimates of Hubble's constant have varied over the decades. (Based on an account compiled by John Huchra of the Harvard-Smithsonian Center for Astrophysics.)

of the programme in 1993, the software had been rewritten to process the observations secured from the unplanned orbit. The positions of 120,000 stars were measured to within one-thousandth of a second of arc and their brightnesses were noted to within 0.2 magnitude, and the Hipparcos Catalogue issued in May 1997 at a symposium in Venice, Italy, reported their distances. By measuring the parallaxes of the 218 stars in the Hyades cluster, and establishing its distance as 151 lightyears, this finally anchored the first 'rung' on the distance scale 'ladder'. In fact, the data was of such quality that Michael Feast at the University of Cape Town in South Africa was able to recalibrate the cepheid period–luminosity relationship, with an uncertainty in the distance of 5 per cent and, finding that they were further away than supposed, he reduced the value of the Hubble constant by 10 per cent, which increased the age of the Universe by a corresponding amount. Furthermore, it was also found that the brightest stars were a little further away than thought, and hence more luminous, and since luminosity is proportional to mass, more massive, and therefore younger. When the stellar models were revised to take this into account, it reduced the ages of the globular clusters to 12 billion years. The pieces of the puzzle were finally beginning to fit together.

KEY PROJECT

Although the Hubble Space Telescope was installed in orbit by the Space Shuttle in April 1990, a flaw in the mirror prevented high-resolution imaging until astronauts fitted a 'corrective optics' package in late 1993. Employing long exposures, it could achieve almost 30th magnitude and give exceedingly sharp pictures of the faint outer regions of galaxies.

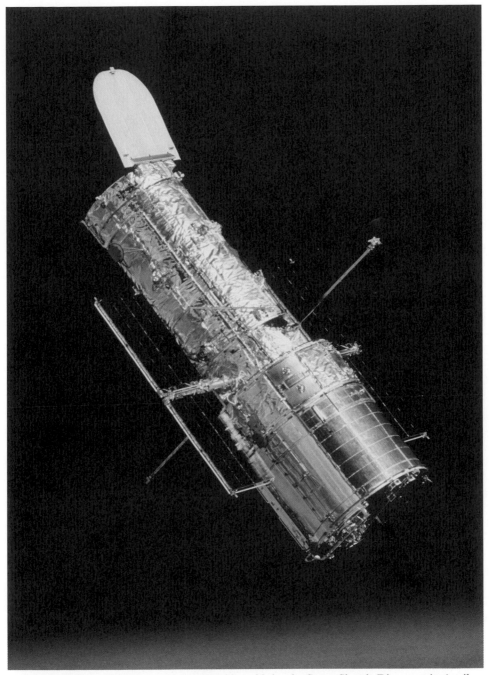

The Hubble Space Telescope was placed in orbit by the Space Shuttle Discovery in April 1990. Although its mirror is only 2.4 metres in diameter, it has the advantage of observing from above the atmosphere.

The goal of the *Key Project on the Extragalactic Distance Scale* was to identify cepheids in galaxies in the Virgo cluster to determine the value of Hubble's constant to within 10 per cent. The 28-member team had three principal investigators: W.L. Freedman of the Observatories of the Carnegie Institution of Washington, located in Pasadena, R.C. Kennicutt at the Steward Observatory in Arizona, and J.R. Mould at the Mount Stromlo and Siding Springs Observatories in Australia. An early target was M100. It was thought to be almost twice as distant as M101 which, Sandage, unable to locate cepheids, had estimated to be 22 million lighyears away on the basis of other indicators. In contrast to Tammann, who had inspected Sandage's plates visually, the Hubble's electronic imagery was computer processed and, as Freedman delightedly found, detecting cepheids was "much easier than we had ever expected". Over the next few years, some 800 cepheids spanning the period–luminosity range were monitored in 18 spirals, supplementing the 13 galaxies in which cepheids were already known.

At 108 million lightyears, NGC4603 in the Centaurus cluster was the most distant spiral investigated by the Hubble Space Telescope Key Project on the Extragalactic Distance Scale. Nevertheless, several dozen cepheids were able to be identified by computer analysis. Courtesy of Jeffrey Newman of the University of California at Berkeley, and NASA.

Table 11.1 Key Project distance measurements

Galaxy		Constellation	Cluster	Type	Million lightyears
NGC3031	M81	Ursa Major	M81 Grp	Sb	1.8 $\pm$ 1.1
NGC3621		Hydra		–	20.6 $\pm$ 2.3
NGC5457	M10	Ursa Major	M101 Grp	Sc	4.1 $\pm$ 2.0
NGC925		Triangulum	NGC925	–	0.3 $\pm$ 2.3
NGC3351	M95	Leo	Leo I	SBb	2.9 $\pm$ 2.9
NGC2090		Columba	–	–	0.0 $\pm$ 2.9
NGC2541		Lynx	NGC2541	–	0.4 $\pm$ 1.9
NGC3319		Ursa Major	–	B(rs)cd	6.6 $\pm$ 1.6
NGC4725		Coma Berenices	NGC4565	–	41.1 $\pm$ 3.3
NGC3198		Ursa Major	–	Sc	47.3 $\pm$ 2.9
NGC7331		Pegasus	NGC7331	Sb	49.2 $\pm$ 3.3
NGC4548	M91	Coma Berenices	Virgo I	SBb	51.8 $\pm$ 6.5
NGC4535		Virgo	Virgo I	S	52.2 $\pm$ 6.2
NGC4321	M100	Coma Berenices	Virgo I	Sc	52.5 $\pm$ 4.2
NGC1365		Fornax	Fornax I	–	59.6 $\pm$ 5.5
NGC1326A		Fornax	Fornax I	–	60.9 $\pm$ 3.9
NGC4414		Coma Berenices	NGC4613	–	62.3 $\pm$ 4.6
NGC1425		Fornax	Fornax I	–	72.4 $\pm$ 3.3

While it was significant to determine accurately the distance to the Virgo cluster, it was only the *closest* rich cluster. It was used to calibrate indicators, so as to estimate distances to more remote galaxies for which the peculiar velocities were insignificant compared to the redshifts. The value of the Hubble constant was then estimated by a variety of methods.

Table 11.2 Key Project estimates of the Hubble constant

Value	Uncertainty		Method
21.48	+ 1.53	–1.84	Surface brightness fluctuations
21.78	+ 0.61	–0.61	Type Ia supernovae
21.78	+ 0.92	–2.15	Tully–Fisher relation
22.08	+ 2.76	–2.15	Type II supernovae
25.15	+ 1.84	–2.76	Fundamental plane

Note: The Tully–Fisher relation was introduced by R.B. Tully and J.R. Fisher in a paper entitled 'A New Method of Determining Distances to Galaxies' published in *Astronomy and Astrophysics* in 1977. It relies on the fact that for spirals there is a relationship between intrinsic luminosity and the rate of rotation, this being determined from the profile of the neutral hydrogen line, which is a distance–independent observable. As refined, this is considered to be a fairly accurate secondary distance indicator.

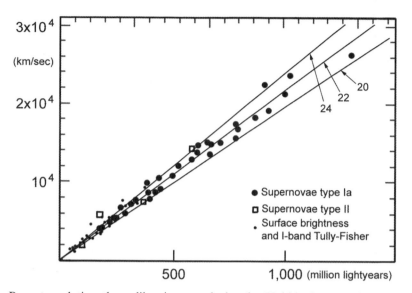

By extrapolating the calibrations made by the Hubble Space Telescope Key Project on the Extragalactic Distance Scale, it is possible to replot recessional velocities (in effect, redshifts) versus distances for more distant galaxies measured using a variety of techniques to determine the value of Hubble's constant. The 'best estimate' was 21.5 ($\pm$2.5) kilometres per second per million lightyears. Courtesy of the Hubble Space Telescope Key Project, 2001.

The 'best estimate' was 21.5 ($\pm$2.5) kilometres per second per million lightyears. For Sandage, who had recently revised his estimate of 15.3 to 17.8 kilometres per second per million lightyears by using Type Ia supernova as standard candles, this result was "a psychological disaster". At least by now there was a slight overlap in the 'error bars' of the rival camps; they were no longer disputing a factor of 2. However, just as it looked as if the issue of the age of the Universe had finally been settled, a surprising discovery was made.

RUNAWAY ACCELERATION!

After graduating from Harvard in 1981, Saul Perlmutter gained his doctorate from the University of California at Berkeley in 1986 and joined the Lawrence Berkeley Laboratory. In 1988 he decided to measure the deceleration parameter, q_0. Most astronomers had given up on this, but Perlmutter, who was a physicist and self-taught astrophysics, thought he had a method of measuring it. He teamed up with Berkeley postdoc Carl Pennypacker and they wrote a proposal. The essence of the measurement was to compare the value of Hubble's constant as it is now with what it used to be. Sandage had attempted to measure departure from the redshift–distance relationship, but had not been able to reach out far enough for this to be unambiguous. Perlmutter estimated that they would need 40 or 50 measurements for galaxies ranging out as far as 7 billion lightyears in order to gain an impression of how the expansion had been slowed. Since supernovae outshine their host galaxies and are simple to spot, they decided to use these as standard candles. They selected Type Ia for their uniformity. These are close binary systems in which a white dwarf accretes hydrogen from its companion until it approaches the Chandrasekhar limit, and as the temperature of its core rises, it undergoes a thermonuclear runaway – a specific physical situation which produces an outburst with a readily recognised light curve. It was essential to detect these on the rise, so that their peak brightness could be determined. Although such outbursts take several weeks to reach their peak, the window of opportunity was nevertheless narrow. Astronomers with experience of seeking supernovae were dismissive of this plan. Fritz Zwicky's study in the 1930s had established that there were only two or three supernovae per galaxy per thousand years – and that included all types. To make the scheme viable, it would be necessary to monitor large numbers of galaxies, and a large telescope would be needed to probe to the requisite 'depth'. To the astronomers, an impracticable task was a waste of valuable telescope time. But Perlmutter had secured some funding. Pennypacker built a special CCD camera to record faint sources over a wide field. Sandage and Tammann had previously demonstrated that a project of this nature could rapidly produce an archive of plates whose analysis could take years. For Perlmutter, time was of the essence. Spotting a supernova was only the first part of this *Supernova Cosmology Project*, the supernova had to be monitored to determine its peak brightness and its redshift had to be measured. This meant that the survey had to be analysed in as near real-time as possible. To this end, Perlmutter had written a computer programme.

By early 1992, after several fruitless observing runs, those who had scorned the idea were saying 'I told you so', and even though the first supernova was spotted in April of that year the project's future was placed under review. However, when the pace of discoveries improved, the funding was safe. One of the severest critics had been Robert Kirshner at the Harvard–Smithsonian Center for Astrophysics, but one of his postdocs, Brian Schmidt (who would soon join the Mount Stromlo and Siding Springs Observatories in Australia) was so motivated by Perlmutter's success that in November 1994 he set up the *High-z Supernova Search Team* in competition, and secured his first supernova six months later.

In 1995 Perlmutter's team was given a few nights on the 4-metre telescope at the Cerro Tololo Interamerican Observatory in the Chilean Andes. The *modus operandi* was for part of the team to exploit a night approaching the 'dark of Moon' to survey 50 to 100 patches of sky, each containing about 1,000 galaxies. Type Ia supernovae take three weeks to reach their peak brightness, so there was a fair chance that when the survey was repeated a month later any supernovae would still be rising. The data was transmitted to Berkeley by the internet and for a day or so the computer would analyse the 100,000+ galaxies. A list of supernova candidates would be e-mailed to Hawaii, where other members of the team would monitor them to measure their peak brightness and use the light grasp of the 10-metre Keck Telescope to take spectra in order to measure their redshifts and verify their Type Ia status. For the most distant cases, the Hubble Space Telescope was called upon for follow-up observations. The years of preparation were paying off, with each observing run producing a number of cases. As team member, Peter Nugent, delightfully put it, the project "was producing supernovae on demand".

Both teams were granted time for one or two observing runs per year, and were clocking up successes. In 1997 Perlmutter achieved his goal of detecting a supernova with a redshift of 83 per cent, corresponding to a range of 7 billion lightyears, and it was feasible to attempt to measure the rate at which the expansion was being slowed. On 24 September 1997, Gerson Goldhaber, an analyst on Perlmutter's team, realised that there was something awry – as the most distant supernovae were fainter than expected, and thus further away. They had made the startling discovery that rather than slowing down, the expansion was *accelerating*. Because they would never live it down if they made an announcement which they later had to retract, they kept silent and checked out each stage of their data processing procedure.

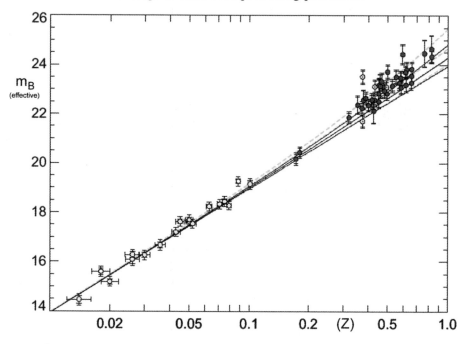

The curvature towards the high end of this plot of redshift (z) versus the effective blue apparent magnitude for Type Ia supernovae implied that rather than slowing down, the rate at which the Universe is expanding is actually *accelerating*. The 18 open circles at low redshift were supplied by a Cálan/Tololo Supernova Survey in 1996. The 42 high-redshift points were measured by the Supernova Cosmology Project. The solid lines show a range of cosmological models with a cosmological constant of zero (upper: $\Omega = 0$ 'open'; middle: $\Omega = 1$ 'flat'; and lower: $\Omega = 2$ 'closed'). The dashed lines are 'flat' models in which Ω steps up from 0 to 1.5 in increments of 0.5 and the cosmological constant runs from 1.0 to –0.5 in increments of –0.5. A statistical analysis indicated a 'flat' Universe with a *positive* cosmological constant driving runaway acceleration. Courtesy of Saul Perlmutter of the Lawrence Berkeley National Laboratory.

Meanwhile, Schmidt's team was undergoing the same agony. They received their first hint that something was amiss in July 1997, and by November, with new data, they were sure of it. Schmidt's reaction when the truth dawned was amazement that this was the result, and horror that it was so unexpected that it would surely be rejected by conservative cosmologists, so he too kept silent. The rivalry between the two teams meant that neither was aware that the other had made the same discovery. In January 1998 Perlmutter made a tentative report to the American Astronomical Society at a meeting in Washington, DC, and Schmidt followed up a month later at a conference in California. As had happened in 1965 with the microwave background, the news was spread by the media before the formal papers appeared in *Science*. In October 1998 the University of Chicago held a symposium to assess the results. Like Schmidt, Perlmutter was shocked by the discovery, but he was amazed by how rapidly cosmologists came to terms with the fact that the Universe was accelerating. This ready acceptance was undoubtedly due to the fact that, at a stroke, acceleration solved so many outstanding questions.

The acceleration meant that the mutual gravitational attraction of the contents of the Universe that acted to slow the expansion was being counteracted by a universal repulsive force. In effect, this was the cosmological constant that Einstein added to general relativity in order to hold the Universe static against collapse. In retrospect, however, despite his dismissal of it as a blunder, it was the *only* term that *could* be added to theory without undermining its symmetry, and so it was not too surprising that it should turn out to have a physical manifestation. It had long been realised that whereas space in general relativity is a continuum which, on the smallest of scales, is 'flat' and therefore has zero energy, in quantum theory 'empty space' seethes with virtual particles. In a sense, therefore, the cosmological constant is an expression of the energy of the vacuum. An energised 'false vacuum' imparts a pressure that expands space. A study of the ripples in the cosmic microwave background gave a measure of the energy of the vacuum and – to general amazement – resolved a long-standing mystery. The visible baryonic matter in galaxies accounts for just a few per cent of the required density for $\Omega = 1$, and the 'dark' non-baryonic matter whose gravity is believed to bind clusters together accounts for only another 30 per cent of the critical value. The 'missing mass', it was now realised, is not present in the form of matter, it is energy, and worse, it is 'hidden' in the vacuum. "This is like a joke," observed Tammann in surprise at the simplicity of the solution, "It's beautiful". As the repulsive force drives the galaxy clusters ever further apart, the rate of expansion will increase – the Universe is undergoing a *runaway* expansion.

In the absence of deceleration, the Hubble Space Telescope team's determination of the Hubble constant as 21.5 kilometres per second per million lightyears made the Universe 14 billion years old. If $\Omega = 1$, as required by the theory of inflation, then (in the absence of acceleration) the actual age, allowing for deceleration, would have been simply two-thirds of this – about 10 billion years. This was in conflict with the age of the globular clusters, even after this was scaled down by the Hipparcos recalibration to 12 billion years. The discovery of acceleration considerably complicated the issue, but Charles Lineweaver of the University of New South Wales in Sidney, Australia, calculated how the cosmological constant influenced the

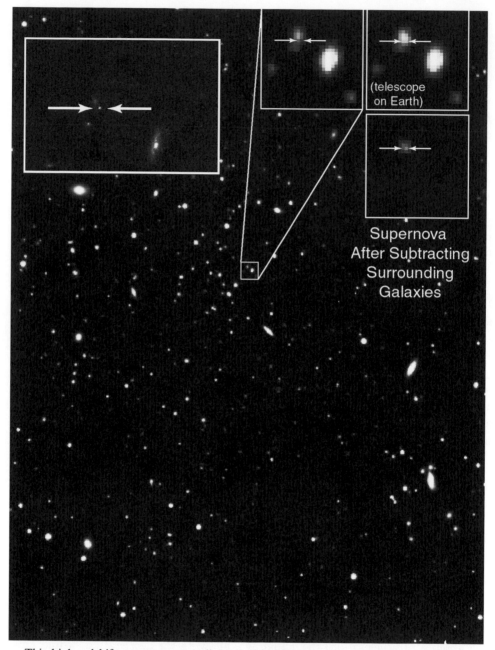

(telescope on Earth)

Supernova
After Subtracting
Surrounding
Galaxies

This high-redshift supernova was discovered by the Supernova Cosmology Project in March 1998. As a follow-up, it was inspected by the Hubble Space Telescope (insert). The immense task facing the team in producing supernovae 'on demand' can be gleaned from the scope of the main image. Courtesy of the Supernova Cosmology Project.

After Type Ia supernovae were found in remote galaxies by the High-z Supernova Search Team in April 1997 using the Canada–France–Hawaii Telescope in Hawaii, the Hubble Space Telescope took these follow-up images. SN1997cj in Ursa Major has a redshift of 50 per cent (left). SN1997ce in Lynx has a redshift of 44 per cent (centre). Both erupted about 5 billion years ago. SN1997ck in Hercules erupted 7.7 billion years ago (right), when the Universe was half its present age. With a redshift of 97 per cent, it took the record for the most distant supernova ever found, but on 15 October 1998 the Supernova Cosmology Project spotted SN1998eq in Pegasus whose 120 per cent redshift indicated that it erupted 10 billion years ago. Courtesy of Peter Garnavich of the Harvard-Smithsonian Center for Astrophysics and the High-z Supernova Search Team.

expansion rate through time. He found that with $\Omega=1$ the rate initially decreased as a result of deceleration when the galaxies were packed closely together, but as the matter thinned out and the vacuum energy began to predominate some 5 billion years ago, the rate picked up. In a paper in *Science* in May 1999 entitled 'A Younger Age for the Universe', he reported that the Big Bang occurred 13.4 billion years ago – an age that provided a comfortable margin for the creation of the stars in the globular clusters. The pieces of the puzzle were indeed beginning to fit together.

FINALLY, $\Omega=1$

In 1986 E.D. Loh and E.J. Spillar at Princeton reported a study in which they used photometry in six optical and near-infrared bands to estimate redshifts for 1,000 galaxies to plot the three-dimensional distribution of *every* galaxy in five narrow

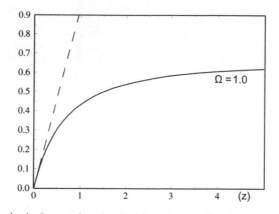

For small cosmological recessional velocities, the relationship with redshift is almost linear. Although a projection to $z=1$ would imply that such objects were receding at the speed of light, this is not so because at a redshift of about 0.25 the relationship diverges. The curve for $\Omega=1$, the situation that is believed to exist, shows how it is possible to have much larger redshifts ($z=5.5$ is the current record).

probes ranging out some 3 billion lightyears, and just penetrating quasar territory. Significantly, this 'core sample' method made no assumptions as to the nature of the gravitating masses, so it provided a measure of the mean density: reporting $\Omega=0.9$ (± 0.3), they concluded that Ω was "indistinguishable from 1.0".

During the first 300,000 years of the Universe, before radiation 'decoupled' from matter, gravitational attraction was overwhelmed by radiation pressure. However, 'dark' (non-baryonic) matter, by not interacting electromagnetically, would not have felt this pressure and so would have been free to move under gravitational attraction. When radiation decoupled, baryonic matter would have responded to this lumpiness. This variation in density would have manifested slight ripples in the temperature of the microwave background, and the value of Ω at that time can be inferred from the angular scale of the ripples, which depends upon the degree of clumping. The ripples noted by the COBE satellite spanned a total range of just 60-millionths of a degree, but its angular resolution was too coarse to estimate Ω. In December 1998, teams led by Kim Cobel of the University of Chicago and Jeff Peterson of Carnegie–Mellon University in Pittsburgh announced studies of small sample areas of the sky from Antarctica which indicated $\Omega=1$. This was confirmed in September 1999 by Lyman Page of Princeton and Mark Devlin of the University of Pennsylvania, who used the Mobile Anisotropy Telescope installed at Cerro Toco, near the town of San Pedro de Atacama in northern Chile. The detector of the 'Boomerang' telescope, which was carried over Antarctica by a stratospheric balloon in 1998, could measure the ripples to an angular resolution 35 times finer than COBE. A.E. Lange of Caltech and Paolo de Bernardis of the University of Rome reported its results in April 2000. Although it mapped only 2.5 per cent of the sky, the ripple scale of slightly less than 1 degree was in excellent agreement with the prediction for $\Omega=1$.

The Microwave Anisotropy Probe that NASA launched in June 2001 was to

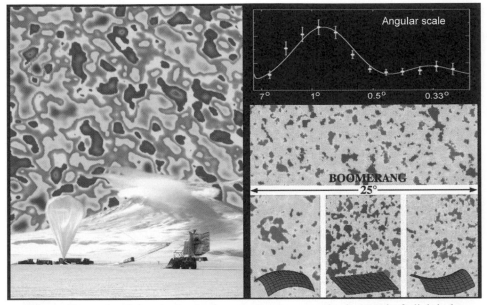

The angular scale of the ripples in the cosmic microwave background of slightly less than 1 degree measured by the Boomerang telescope on a stratospheric balloon over Antarctica supported the case for the Universe being 'flat'. The depiction set against Mount Erebus gives an impression of how the background would appear if we were able to see it with the naked eye. Courtesy of the Boomerang Project.

perform an *all-sky* survey at sufficiently high resolution to measure these ripples, and so determine the value of Ω in the early Universe. The project was conceived by D.T. Wilkinson who had worked on Dicke's radiometry team in 1965, and remained at Princeton. Upon Wilkinson's death in September 2002, the spacecraft was renamed in his honour. The preliminary results, reported in February 2003, proved that $\Omega = 1$; that, to within the unprecedented accuracy of 1 per cent, the Big Bang occurred 13.7 billion years ago; and that the Universe comprises 4 per cent ordinary matter, 23 per cent of an unknown type of dark matter, and 73 per cent of an energy field that acts to accelerate the rate at which space is expanding.

While it might be fair to say that we now have the factual version of the story, we might be wise to recall that astronomers of the early twentieth century also felt confident that they had settled the important issues.

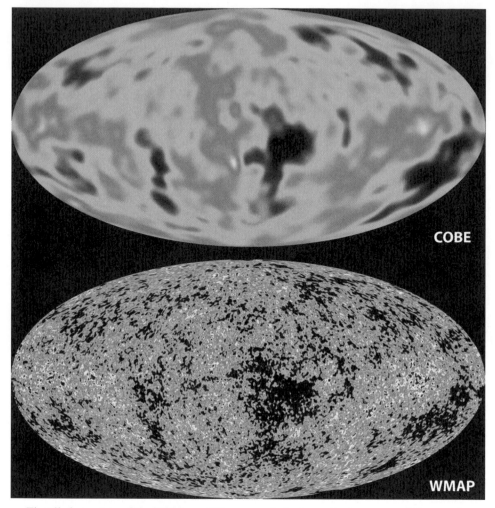

The all-sky survey of the Wilkinson Microwave Anisotropy Probe measured the angular scale of the ripples in the cosmic microwave background some 380,000 years after the Big Bang, bringing the COBE survey into sharp focus. Courtesy of D.N. Spergel of Princeton University, the WMAP Science Team and NASA.

12

The Big Bang

THE NATURE OF BLACK HOLES

In 1959 S.W. Hawking went to Oxford to study physics. After graduating in 1962 he went to Cambridge. In early 1963, soon after his 21st birthday, he began to develop the symptoms of motor neurone disease, a sclerosis that reduces voluntary muscle control; he was told that it was progressive and incurable. After a frustrating few years of awaiting imminent death Hawking, at the prompting of his supervisor, Dennis Sciama, a theoretical cosmologist favouring the steady state theory, attended a lecture on singularities by Roger Penrose at the University of London and, suitably reinvigorated, he resumed his research. Singularities had been investigated in terms of collapse, but Hawking envisaged the creation of the Universe as a singularity, thereby offering an alternative mechanism for Gamow's primordial fireball. Upon graduating, Hawking remained at Cambridge, and although he was wheelchair-bound by 1969 his intellect thrived and he started to see black holes in a new light. In November 1970, he recognised that they exhibited behaviour which was analogous to thermodynamic properties: specifically, because the radius of the event horizon was proportional to the black hole's mass, and could only increase as material was consumed, the size of the event horizon was analogous to *entropy*, which also inexorably increased for a closed system.

In 1972 Jakob Bekenstein, a student of J.A. Wheeler at Princeton, reconsidered the fate of the entropy of the mass which fell into a black hole. The traditional view was that this 'disappeared'. Bekenstein argued in his doctoral thesis that black holes did not simply exhibit behaviour *analogous* to entropy, they *had* entropy, and lots of it, as they were the 'end product' of stellar evolution. The increase in 'disorder' as the 'information' in the infalling material 'disappeared' increased the entropy of the black hole. Bekenstein argued that the area of the event horizon was *a measure* of the hole's entropy. The significance of this assertion was that a black hole now had a property that was characteristic of the quantum realm – this was the first direct link between general relativity and quantum mechanics. Bekenstein's argument was based on Wheeler's saying that "black holes have no hair", by which he meant that they displayed no trace of what they had consumed, they all display the same 'hairstyle'. It had been thought that they had only mass, angular momentum and electric charge, but it now appeared that they also possessed entropy. Bekenstein boldly published his conclusions in 1973 in a paper 'Black Hole Thermodynamics'. However, Hawking was sceptical, because if thermodynamics applied to black holes they would require to possess a *temperature* (the value of which would be determined by the strength of the gravitational field at the event horizon) and a non-zero temperature would mean that they *radiated*, whereas everyone knew that *nothing* could escape a black hole. Hawking argued that thermodynamical analogies were useful mathematical tools for studying black holes, but should not be taken literally because thermodynamics did not really apply to an object having a temperature of absolute zero. Nevertheless, in 1971 Hawking had speculated that even though a substellar mass would not collapse under its own gravitational attraction, minor irregularities in the extremely dense primordial fireball during its first 10^{-10} second could have squeezed 'lumps' into singularities, making 'primordial black holes' of *arbitrarily small* masses. With student Bernard Carr, Hawking investigated further. The event horizon of a black hole of 1 solar mass is some 6 kilometres in diameter, one of 50 million solar masses is about 1 AU, and one of 10^8 solar masses is 10^9 kilometres, which is comparable to the diameter of the Solar System. In contrast, the event horizon of a black hole of 1 billion tonnes would be 10^{-15} metre, making it the size of a proton! On such a scale, quantum mechanics would have to be taken into account. In November 1973 Hawking began to ponder whether an electron might settle into orbit around such a black hole, retained by the intense gravitational gradient rather than electrostatic attraction. To his amazement, he found that the Uncertainty Principle required such a black hole to *radiate* because the intense gravitational gradient *at* the event horizon would tend to draw in a newly created *virtual* particle which was moving in the direction of the hole, while the other member of the pair, with outwardly-directed kinetic energy, would move away from the event horizon and emerge as a *real* particle. Black holes were *not* totally black! They glow with a black body spectrum with a temperature inversely proportional to their mass. While the temperature of a stellar-mass black hole would be only 10^{-7} K, and a supermassive black hole in a galaxy's core would be even chillier, a primordial black hole of 1 billion tonnes would have a temperature of 10^{11} K and would radiate gamma rays. As it lost energy, its mass would decrease and its event horizon would shrink, which would intensify the gravitational gradient and promote increased radiation in a runaway 'evaporation'. When Hawking tentatively presented this discovery at a conference at Oxford in February 1974, the moderator, John Taylor, a theorist at the University of London, dismissed it as "absolute rubbish". Hawking had entitled his paper 'Black Hole Explosions?', with the question mark providing him with an escape in the event that somebody immediately spotted a flaw in his argument. Even though his argument proved robust, it was several years before people accepted the viability of 'Hawking radiation'. The lifetime of a black hole is a linear function of its mass: one of 1 million tonnes will evaporate in about 30 years, and one of 1 billion tonnes will last 300 million years. All of the very-low-mass primordial black holes must therefore already have evaporated. Very large black holes will be radiating exceedingly slowly. At any given time, those of a specific mass will be approaching the runaway phase. Nowadays, ones of around 5 billion tonnes are due to evaporate. As they do so, their radiation will run up the spectrum and end with an intense blast of gamma radiation of about one-tenth of a second's duration. Intriguingly, flashes of gamma rays *have* been observed from random directions in space, but research suggests that they are massive black holes in the process of formation rather than small ones in the process of evaporation.

NO SINGULARITIES!

The realm of the very large is described by general relativity, and the realm of the very small is described by quantum mechanics. Although mutually incompatible, we need both to describe *very compact massive objects*. The concept of a *singularity* is a consequence of general relativity, but general relativity's premise that space is a smooth continuum is inconsistent with the seething vacuum inherent in the quantum realm. This incompatibility was resolved by string theory, which applies at Planck energy and actually *requires* gravitation. As a result of treating space as a continuum that can be divided ever more finely, general relativity permits space to be curved to such an extent that a point-like 'flaw' forms in space. String theory does not permit this. In the late 1980s, in exploring the theory's cosmological implications, Robert Brandenberger at Brown University in Providence, Rhode Island, and Cumrun Vafa at Harvard realised that although the temperature of the fireball rose as they reversed time in their computer, the temperature reached a *maximum* as the curling of the spatial and temporal dimensions approached the Planck scale. They reasoned that because a string cannot be forced into a *smaller* volume than this, the Universe could *not* have begun as a point of infinite density and an infinite temperature. As the smallest object in string theory is the perfectly curled multi-dimensional *Planckian nugget*, there is no such thing as a singularity. In much the same manner as degenerate electrons provide the pressure to inhibit the collapse of a white dwarf and degenerate neutrons inhibit the collapse of a neutron star, string theory inhibits the creation of a singularity – the dramatically increasing energy of the collapse is transferred into the string. D.J. Gross and Paul Mende at Princeton discovered in 1988 that an energised string *expands*. As a black hole forms, instead of leaving a singularlity it energises a string to many times the Planck energy, causing it to *grow*, possibly in an independent set of dimensions. Lee Smolin, a student of Sidney Coleman's at Harvard who gained his doctorate in 1979, has speculated that *another* space-time will be spawned which is linked to our own only by what we see as a black hole. If so, then the profusion of black holes in our Universe indicates a multiplicity of isolated Universes. And of course, this raises the intriguing question of whether what we think of as the Big Bang is really no more than an *internal perspective* of the formation of a black hole?

THE COOLING FIREBALL

String theory implies that instead of beginning as a point of infinite density, the Universe formed as a *Planckian nugget* with many dimensions curled up in a highly symmetrical state. Three dimensions uncurled to establish the space with which we are familiar, and another uncurled to form what we think of as the time axis. As the smallest measurable distance is 10^{-35} metre, and since light can travel this distance in 10^{-43} second, this is the *smallest measurable unit of time*. Because 'time zero' has no meaning, the story of the Big Bang starts at this point.

   As gravitation is a measure of the curvature of space, the loss of symmetry as the spatial dimensions started to uncurl *introduced* the force of gravitation. This also marked the loss of symmetry between fermions and bosons. With an initial energy density of about 10^{19} GeV, pair production was rampant and the fireball was a sea of 'lepto-quarks' and associated bosons. As the temperature decreased from 10^{28} to 10^{26} K during the interval from 10^{-35} to 10^{-33} second, the spontaneous breaking of a symmetry drew distinctions between quarks and leptons and split the colour and electroweak forces. As the energy density fell below that for pair production of the X-bosons, they ceased to be created and those which existed either annihilated one another or promptly decayed. In that early era, the profile of the total energy versus the strength of the Higgs field was U-shaped, and the minimum energy state had a zero field strength; the *net value* of the Higgs field was zero.* As the profile switched to a W-shape during the phase change, the transfer of kinetic energy to the field gave the minimum energy state a non-zero potential. The Higgs field became manifest in the same way that a magnetic field is 'frozen' into cooling iron. Although the *net* value of the field prior to the phase change was zero, its *instantaneous value* would have been fluctuating. When the profile became W-shaped the net energy of the field was zero but the *overall energy* was no longer zero – the Universe had assumed a false vacuum. As long as this metastable state persisted, the fireball would have been dominated by the field. Einstein's equations of general relativity oblige a Universe in which most of the energy is locked in a field to undergo exponential expansion, with its radius of curvature doubling and redoubling in a certain time interval. Depending on the specific conditions, there may have been time for the Universe to undergo *100 cycles* of doubling. Whereas this would have increased the radius by a factor of about 10^{30}, extrapolating the rate of expansion observed today back over that same interval would have produced only a ten-fold increase. Such an extrapolation indicates that the observable Universe was about 10 centimetres across at the *end* of this period of inflation. The volume from which it inflated was therefore considerably smaller than that of a proton – although, of course, there were no protons then. Lest exponential growth should give the impression that inflation violated Einstein's speed-of-light rule, the rule limits motion *through* space, not the expansion of space itself. On the question of why there were such symmetry-breaking

---

*   In fact, there were several Higgs fields, but the sum of their potentials can also be represented by a single value.

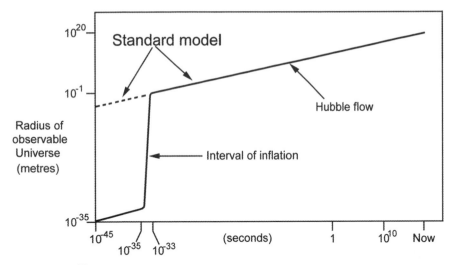

At about 10^{-35} second after the Big Bang, a spasm of exponential expansion lasting until 10^{-33} second inflated the radius of the Universe by a factor of 10^{30}, smoothing out the various energy fields and flattening space to make $\Omega = 1$.

scalar fields as the Higgs extant in the early fireball, they are found to be a *feature* of string theory.

As a vector field, electromagetism was 'stretched' by inflation which reduced its energy density, with resultant cooling. Although the gradients of the scalar fields were smoothed out during inflation, their potentials remained constant and their total energy increased in proportion to the cube of the increasing radius. The energy of the Higgs field would therefore have been enormously *enhanced* by inflation. When the field settled into the true vacuum, all this energy was shed by being converted into another form, in much the same manner as the kinetic energy of cooling iron atoms is converted into a magnetic field. A physical analogy is an alpha particle tunnelling from a nucleus and flying off at high speed as its potential energy is transformed into kinetic energy. As the Higgs field energy 'condensed', it bestowed mass on particles, which could not be released by further cooling. As soon as the fireball ceased to be dominated by the energy in the Higgs field, the exponential expansion halted. The reason we do not perceive the non-zero potential of this field in its true vacuum state is that the world-view of a hypothetical magnetic creature living in a magnet would be dictated by the fact that the magnetic field bestowed a preferred axis. The creature would find it difficult to imagine the removal of the field and the reinstatement of a symmetry in which there was no directionality. It is similar for us. We perceive the prevailing directionality of the Higgs field as mass. It is difficult to step back and imagine its absence in every day 'common sense' terms, but such conditions can be explored mathematically. At the end of the inflationary period, therefore, the Universe was *vastly* larger than it had been beforehand, and it was loaded with mass.

As the energy density continued to decline, the coupling constants of the forces diverged because collisions became progressively less able to sense a particle's bare charge. Also, while the strength of the electroweak force declined as the efficiency of 'shielding' by vacuum polarisation increased, that of the colour force spectacularly increased as the quarks became immersed in clouds of the self-interacting gluons. At 10^{-10} second, by which time the temperature was reduced to 10^{13} K, the electroweak symmetry spontaneously broke as another Higgs field froze, but because the energy density was now a mere 100 GeV and the fireball was full of matter, there was no significant inflationary effect. Prior to this, pair production of the electroweak bosons had been rife, but this ceased, and those that existed either annihilated each other or decayed. It was this phase change that split the electroweak force into its electromagnetic and weak components. Within a *billionth* of a second, therefore, the initial state of pure symmetry had been progressively broken and the forces with which we are familiar had become manifest.

At about 10^{-5} second, by which time the fireball had cooled to an energy density of 1 GeV, the quarks started to 'condense' into the hadrons, within which they have since been confined. Although the distinction between quarks and leptons had been drawn at the end of grand unification and the conservation of lepton number became applicable at that time, the conservation of baryon number remained latent until the baryons were created at the start of the hadron era. As an energy density of 1 GeV is an order of magnitude greater than that in the core of a hydrogen-burning star, any aggregates of baryons that might have formed at this time would promptly have been broken up again. As the quarks condensed, any that were slow to find mates would have drifted through a sea of hadrons that could not accept gate-crashers. If one met an antiquark, they would have been able to combine as a meson. Is it possible that a few are *still* adrift? When Gell-Mann devised the idea of quarks, he had balked because there was no evidence for their fractional electric charges. His critics had cited this absence of evidence as evidence of absence. The eventual success of quantum chromodynamics verified the quark theory, and the asymptotic freedom that prevents quarks from being drawn out of hadrons explained the absence of evidence of fractional charges. Indeed, the absence of evidence for fractional charges was cited as evidence of the absence of *free* quarks. Nevertheless, because the observation of a fractional charge could not be explained in any other way, in 1969 William Fairbank of Stanford's Low Temperature Laboratory utilised a magnetic field to levitate a niobium sphere within a cryogenically chilled apparatus and was able to isolate a residual fractional charge of $-1/3$. After refining the experiment to eliminate sources of error, and finding fractional charges of $\pm 1/3$ and $\pm 2/3$, Fairbank reported his result in 1977. Although other researchers have not been able to verify this experiment, the prospect is intriguing because free quarks would possess 'naked' colour.

THE MATTER–ANTIMATTER IMBALANCE

In the early 1960s, H.O.G. Alfvén, professor of plasma physics in Stockholm, Sweden, postulated that there were equal amounts of matter and antimatter. He also argued that because the distribution was not uniform, annihilation occurred preferentially at the interfaces between regions, with the radiation pressure of the resulting gamma rays sweeping the interfaces clean of material. This idea was popular until detectors sent aloft on rockets failed to see the characteristic radiation from the annihilation of material that strayed into the interfaces.

After the Second World War, A.D. Sakharov, who graduated from Moscow State University in 1942, worked on the hydrogen bomb. In 1953, aged 32, he became the youngest-ever member of the Soviet Academy of Sciences. In the 1960s his research focused on the Big Bang. In 1965 Peebles reaffirmed the conclusion of Gamow's team that the primordial fireball was too hot for atomic nuclei to form, implying that the Big Bang created subatomic particles. Dissatisfied with Alfvén's theory of matter and antimatter regions, Sakharov suggested in 1967 that only matter had emerged from the Big Bang. For such a matter–antimatter imbalance to have arisen, the fireball must have begun in a state of thermal equilibrium with a balance between the processes which turned radiation into particles (pair production) and turned pairs of particles into radiation (annihilation). Initially, the reactions that created assemblages of particles would have been balanced by those that dissociated the assemblages into their constituents, but as the temperature fell, this equilibrium would have favoured the lower-energy state. At that time, Murray Gell-Mann had proposed that hadrons were composed of quarks, but there was no evidence for this. As we now know, the hadrons (namely mesons and baryons) formed in a phase change at the end of the era of quarks, and a succession of phase changes gave rise to more sophisticated aggregates. Sakharov was considering the phase change in which protons and neutrons (a subset of the baryons) formed nuclei. The processes of pair production and annihilation both involve particle–antiparticle pairs, and if these had been the only processes at work then the fireball would have left only the cosmic background radiation field. To have left a *residue* of particles, there must also have been processes that turned particles into other particles. Sakharov also pointed out that some processes must have made baryons out of non-baryons (which meant that lepton and baryon numbers were not conserved), and to build the matter–antimatter imbalance baryon interactions must have violated both the C and CP symmetries. The fact that the fireball was cooling gave a directionality to time. The effect of violating CP could be compensated for by violating T (which is why matter travels 'forward' in time) to uphold CPT.

Because both baryon and lepton numbers were believed to be conserved when Sakharov proposed this theory, his paper attracted little attention. However, grand unification *required* baryonic matter to be unstable and to decay into leptons, which implied that these were not fundamental quantities. In 1978 Motohiko Yoshimura at Tokyo University studied the properties of the X-bosons of grand unification and (essentially rediscovering Sakharov's chain of reasoning) realised that they would have left an excess of quarks over antiquarks. For example, by emitting an X-boson, a 'u' quark can transform into the corresponding antiquark. If the X-boson were to be absorbed by a nearby 'd' quark, this would turn into an antielectron. If the X-boson did not find another quark to interact with, it could decay into a 'd' antiquark and an antielectron. Hence, in the era of grand unification, the X-bosons could

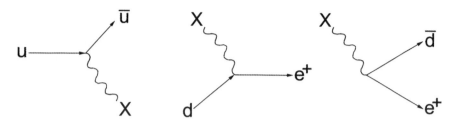

If a 'u' quark emitted an X-boson during proton decay in the era of fermion unification, the X-boson might transform a nearby 'd' quark into an antielectron, but if it failed to find another quark with which to interact, it could decay into a 'd' antiquark and an antielectron. Since the X-bosons could create quarks and leptons and turn quarks into leptons, they would have left an excess of quarks over antiquarks at the end of the era of fermion unification.

both create quarks and leptons and turn quarks into leptons. By violating the component symmetries of CPT, such interactions accumulated an excess of quarks over antiquarks.

In the era of hadrons, there was an evolving equilibrium between the processes of pair production and annihilation. By about $T+10$ seconds, the energy density had decreased so that photons no longer had the energy for pair production to keep pace with annihilation. This was a *gradual* transition, because the departure from thermal equilibrium affected the more massive particles first. An intense burst of annihilation eliminated almost all of the mass from the fireball, transforming it back into energy in the form of photons at their annihilation energies, eliminating all of the antimatter and leaving a residue of matter.

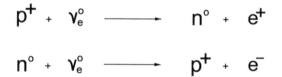

The neutrino interactions which turn protons into neutrons and back again.

Early in the hadron era, the neutrinos were sufficiently energetic to turn protons into neutrons and back again in a reversible reaction. Being massless, the energy of a neutrino is all in kinetic form. As long as the temperature was sufficiently hot for the collisional energy density to make up the mass differences, the rates of the reactions that turned protons into neutrons and back again were in balance. However, because the neutron is slightly more massive than the proton, it took slightly *more energetic* neutrinos to turn protons into neutrons than it did for the reverse. By about $T+1$ minute, the energy density had fallen below the level for neutrinos to supply the mass–energy difference to turn protons into neutrons, and as this reaction ceased a baryonic imbalance developed in favour of protons. By $T+3$ minutes, there were six times as many protons as neutrons and the temperature was cool enough for protons and neutrons to form nuclei. This primordial nucleosynthesis produced nuclei of deuterium (a proton paired with a neutron), tritium (a proton and two neutrons), helium-3 (two protons and a neutron) and helium-4 (two of each). Although this process began as the temperature fell below 10^9 K, the nuclei of different atomic masses became stable at different temperatures. By $T+13$ minutes, most of the remaining free neutrons had spontaneously decayed into protons. Soon the plasma had cooled sufficiently to prohibit electron–antielectron pair production. As the fireball's energy density fell below its matter density, it ceased to be dominated by radiation.

Producing a plasma from gas does not occur at a specific temperature (such as occurs in boiling a fluid) because there are *degrees* of ionisation. Depending upon the element, a temperature of a few thousand degrees will liberate an electron from an atom to produce a plasma, but millions of degrees are required for *total ionisation* in which all nuclei are bare. When the Universe cooled, the same occurred in reverse: it took a long time for the shells to fill up to form neutral atoms. During the transition, which lasted for several hundred thousand years, the fireball ceased to be radiation dominated and gradually became a sea of neutral atoms. At the start of the plasma era, when the temperature was a billion degrees, the fireball was still hot enough to be in thermodynamic equilibrium and the photons, which were mostly in the form of X-rays and gamma rays, could not travel very far before interacting with matter. The radiation pressure of photons scattering off the profusion of free electrons inhibited the formation of large-scale structures by gravitational contraction. (To be strictly accurate, it prevented *baryonic* matter from condensing; non-baryonic matter, which does not interact electromagnetically, would not have felt this pressure and would thus have been free to condense.) When most of the electrons were captured by nuclei to form neutral atoms, the mean free path of the photons became near-infinite and so as the radiation field 'decoupled' from matter the Universe was rendered transparent. The cosmic microwave background observable today is a relic of this plasma, redshifted to a temperature equivalent to a 3 K black body. There were density ripples in the fireball when the radiation decoupled (as revealed by the COBE data). Baryonic matter would have been attracted to the density ripples of non-baryonic matter, and galaxies formed where baryonic matter concentrated.

THE EARLIEST GALAXIES

In 1994, with the Hubble Space Telescope's optics fixed, projects requiring very-high-resolution imagery of remote galaxies could finally be undertaken. Even with its restored light grasp, the telescope required very long exposures to record such faint objects. To study the morphology of early galaxies, three clusters were selected at a range of distances. In the 'near' cluster, some 5 billion lightyears away, there was a surprisingly high percentage of spirals, many of which were interacting, and a large number of fragments torn off in collisions. Interactions were evidently commonplace when galaxies were densely packed. Although the fragments were small, shockwaves

For 10 days in December 1995, the Hubble Space Telescope stared at a seemingly empty patch of sky in Ursa Major for this 'Deep Field' image, which revealed that the early Universe was surprisingly rich in galaxies. Courtesy of NASA/STScI.

in the hydrogen clouds prompted intense bursts of star formation which made them very luminous. The only recognisable galaxies in the cluster 9 billion lightyears away were ellipticals, which was a surprise, but there were many fragments dubbed 'blue dwarfs' because they radiated intensely in the blue despite being heavily redshifted. There were only a few ellipticals in the third cluster, 12 billion lightyears away, but it also contained a quasar.

In 1995, for its Deep Field Project, the Hubble Space Telescope integrated a tiny patch in Ursa Major for 120 hours (in fact, this was built up by 342 exposures taken over a 10-day period). This area had three virtues: (1) it was out of the galactic plane and so was clear of the obscuration that limits our view near the plane of the Milky Way; (2) it was uncluttered by stars; and, most crucially, (3) it did not overlap any known cluster. The objective was to peer as far into space as possible to find out, at last, what the faintest, most remote, and hence the earliest galaxies, looked like. The result was an astonishingly rich field of 2,000 objects, many of which were irregular 'clumps' of stars. Analysis revealed these to be gravitationally bound into groups destined to coalesce into galaxies. In 1998 the Hubble Space Telescope made a similar survey of a patch of sky in Tucana, as its Southern Deep Field. As part of a Faint Infrared Extragalactic Survey, one of the 8.2-metre elements of the Very Large Telescope of the European Southern Observatory integrated for 100 hours in 2002 to secure near-infrared images of the southern field at 2.3 microns, a wavelength at which the Hubble Space Telescope was uncompetitive. The light from these galaxies was emitted when the Universe was less than 2 billion years old. As the visible light has been redshifted into the near-infrared, these images are more readily compared with nearer galaxies at visual wavelengths. Despite their youth, some of the galaxies had spiral structure. "These results demonstrate that very deep observations in the near-infrared are essential to obtain a proper census of the earliest phases of the universe," pointed out Marijn Franx from the University of Leiden, the leader of the team. At long last, astronomers were seeing the stars that formed at the end of the so-called 'dark age' of the early Universe. The primary task of the 7-metre James Webb Space Telescope which NASA intends to launch in 2010 will be to investigate these proto-galactic objects.

DIM PROSPECTS

What fate awaits our Universe? As the rate of expansion accelerates, the galaxies will race off into oblivion, their light being redshifted to insignificance until only the gravitationally bound members of the local supercluster remain within our 'horizon'. Long after the stars have burned out and all is dark, the ashes will be lit from time to time by bursts of gamma rays as a black hole of stellar mass evaporates. However, it will be 10^{67} years before the black hole of Cygnus X-1 evaporates, and by the time the supermassive holes in the cores of galaxies erupt it will take an eternity for their radiation to reach even their nearest neighbours, during which passage it may well be severely redshifted by the ever-increasing rate of expansion. And if, as seems likely, baryonic matter is unstable, the Universe would appear to be destined to decay into

In October 1998 the Hubble Space Telescope took its 'Deep Field South' image in the constellation Tucana, near the south celestial pole (a section of which is shown here). The observed light was emitted in the ultraviolet. Courtesy of NASA/STScI.

In 2002 the European Southern Observatory's Very Large Telescope documented the 'Deep Field South' at 2.3 microns to record the visible emission which had been redshifted down into the near infrared. Courtesy of ISAAC/ANTU/VLT/ESO.

a tenuous sea of leptons and chilly radiation, and it will, in reality, have been a waste of time. As Steven Weinberg put it, "the more the Universe seems comprehensible, the more it also seems pointless".

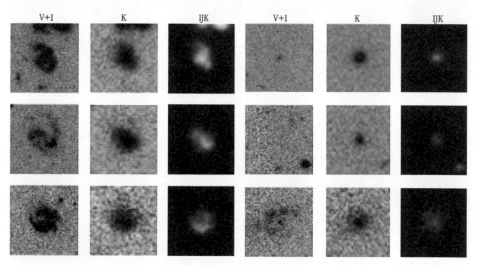

Although the galaxies in the Very Large Telescope's infrared imagery of the 'Deep Field South' are less than 2 billion years old, there is evidence of spiral structure. Courtesy of ISAAC/ANTU/VLT/ESO and HST/WFPC2/NASA/STScI.

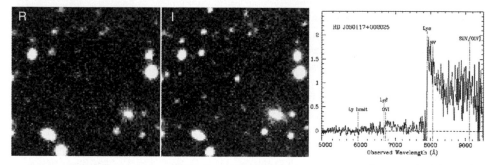

Quasar RD300 was discovered in January 2000 during a survey of extremely faint red objects by a team of astronomers headed by Daniel Stern of the NASA/Caltech Jet Propulsion Laboratory in Pasadena using a variety of telescopes. Due to its record-setting redshift $z = 5.5$, it is evident only in the I-band (far red) image, and not in the R-band (red) image. Because the prominent Lyman-alpha emission in the ultraviolet at 1,200 Ångströms has been shifted down to 8,000 Ångströms in the infrared, it is the faintest known quasar in the visual range. The images were taken with the 4-metre Mayall Telescope of the Kitt Peak National Observatory and the spectrum with one of the 10-metre Keck Telescopes in Hawaii.

Further reading

A popular history of astronomy during the nineteenth century, Agnes M. Clerke, Adam & Charles Black, 1885.

The expanding universe, A.S. Eddington, Cambridge University Press, 1933.

Introduction to quantum mechanics, Linus Pauling and E. Bright Wilson, McGraw-Hill, 1935.

The nature of the physical world, A.S. Eddington, Dent/Everyman, 1935.

The evolution of physics: from early concepts to relativity and quanta, Albert Einstein and Leopold Infeld, Touchstone/Simon & Schuster, 1938.

A concise history of astronomy, Peter Doig, Chapman & Hall, 1950.

The nature of the universe: a series of broadcast lectures, Fred Hoyle, Blackwell, 1950.

The milky way, Bart J. Bok and Priscilla F. Bok, Harvard University Press, 1957.

The realm of the nebulae, Edwin P. Hubble, Dover Publications, 1958.

Ultimate particles of matter, D.T. Lewis, Chantry Publications, 1959.

And there was light: the discovery of the universe, Rudolf Thiel, Readers Union, Andre Deutsch, 1959.

The neutron story, Donald J. Hughes, Heinemann, 1960.

Astronomy of the 20th century, Otto Struve and Velta Zebergs, Macmillan, 1962.

Elementary particles: a short history of some discoveries in atomic physics, Chen Ning Yang, Princeton University Press, 1962.

The world of elementary particles, Kenneth W. Ford, Blaisdell Publishing, 1963.

The atomic nucleus, M. Korsunsky, Dover, 1963.

The discovery of the electron: the development of the atomic concept of electricity, David L. Anderson, Van Nostrand, 1964.

Explaining the atom, Selig Hecht, The Scientific Book Club, 1964.

The structure of the atomic nucleus, J. Kvasnica, Iliffe Books, 1967.

Their majesties' astronomers: a survey of astronomy in Britain between the two Elizabeths, Colin A. Roman, The Bodley Head, 1967.

Atomic and nuclear physics, T.A. Littlefeld and N. Thorley, Van Nostrand Reinhold, 1968.

Violent universe: an eye-witness account of the commotion in astronomy 1968–69, Nigel Calder, BBC, 1969.

The discovery of our galaxy, Charles A. Whitney, Angus & Robertson, 1971.

Black holes: the end of the universe?, John Taylor, Souvenir Press, 1973.

The Einstein decade: 1905–1915, Cornelius Lanczos, Scientific Books, 1974.

Asimov's biographical encyclopedia of science and technology, Isaac Asimov, Pan, 1975.

Einstein, Benesh Hoffmann, Paladin, 1975.

Exploring the galaxies, Simon Mitton, Faber & Faber, 1976.

The universe: its beginning and end, Lloyd Motz, Abacus, 1977.

The key to the universe: a report on the new physics, Nigel Calder, BBC, 1977.

The collapsing universe: the story of black holes, Isaac Asimov, Hutchinson, 1977.

Space and time in the modern universe, P.C.W. Davies, Cambridge University Press, 1977.

The ambidextrous universe: mirror asymmetry and time-reversed worlds, Martin Gardner, Charles Scribner, 1978.

The first three minutes: a modern view of the origin of the universe, Steven Weinberg, Fontana, 1978.

Eyes on the universe: a history of the universe, Isaac Asimov, Quartet, 1978.

Einstein's universe, Nigel Calder, BBC, 1979.

In the centre of immensities, Bernard Lovell, Hutchinson, 1979.

The red limit: the search for the edge of the universe, Timothy Ferris, Corgi, 1979.

Black holes and warped spacetime, William J. Kauffmann, Bantam, 1980.

From atoms to quarks: an introduction to the strange world of particle physics, James S. Trefil, Charles Scribner, 1980.

The nature of matter, J.H. Mulvey (Ed.), Clarendon Press, 1981.

X-ray astronomy, J. Leonard Culhane and Peter W. Sanford, Faber & Faber, 1981.

The accidental universe, P.C.W. Davies, Cambridge University Press, 1982.

The cosmic onion: quarks and the nature of the universe, Frank Close, Heinemann, 1983.

The quest for quarks, Brian McCusker, Cambridge University Press, 1983.

God and the new physics, P.C.W. Davies, J.M. Dent, 1983.

The edge of infinity: beyond the black hole: naked singularities and the destruction of spacetime, P.C.W. Davies, Oxford University Press, 1983.

The creation of matter: the universe from beginning to end, Harald Fritzsch, Basic Books, 1984.

The moment of creation: Big Bang physics from before the first millisecond to the present universe, James S. Trefil, Collier/Macmillan, 1984.

The cosmic code: quantum physics as the language of nature, Heinz R. Pagels, Pelican, 1984.

Quantum mechanics, P.C.W. Davies, Routledge & Kegan Paul, 1984.

Quarks: the stuff of matter, Harald Fritzsch, Pelican, 1984.

Superforce: the search for a grand unified theory of nature, P.C.W. Davies, Counterpoint/Unwin, 1985.

QED: the strange theory of light and matter, Richard P. Feynman, Penguin, 1985.

The particle connection, Christine Sutton, Touchstone/Simon & Schuster, 1985.

The left hand of creation: the origin and evolution of the expanding universe, John D. Barrow and Joseph Silk, Counterpoint/Unwin, 1985.

Perfect symmetry: the search for the beginning of time, Heinz R. Pagels, Simon & Schuster, 1985.

Quarks: frontiers in elementary particle physics, Y. Nambu, World Scientific, 1985.

The cosmic inquirers: modern telescopes and their makers, Wallace Tucker and Karen Tucker, Harvard University Press, 1986.

The forces of nature, P.C.W. Davies, Cambridge University Press, 1986.

The particle hunters, Yuval Ne'eman and Yorak Kirsh, Cambridge University Press, 1986.

In search of the Big Bang: quantum physics and cosmology, John Gribbin, Heinemann, 1986.

The great design: particles, fields and creation, Robert K. Adair, Oxford University Press, 1987.

Elementary particle physics, I.R. Kenyon, Routledge & Kegan Paul, 1987.

The hunting of the quark: a true story of modern physics, Michael Riordan, Touchstone/Simon & Schuster, 1987.

The particle explosion, Frank Close, Michael Marten and Christine Sutton, Oxford University Press, 1987.

Asimov's new guide to science, Isaac Asimov, Penguin, 1987.

The omega point: the search for the missing mass and the ultimate fate of the universe, John Gribbin, Heinemann, 1987.

The quantum universe, Tony Hey and Patrick Walters, Cambridge University Press, 1987.

Discovering the universe, William J. Kauffmann, W.H. Freeman, 1987.

Superstrings: a theory of everything?, P.C.W. Davies and J. Brown (Eds.), Cambridge University Press, 1988.

The dark side of the universe: a scientist explores the mysteries of the cosmos, James S. Trefil, Charles Scribner, 1988.

A brief history of time, Stephen Hawking, Bantam, 1988.

Superstrings and the search for the theory of everything, F. David Peat, Contemporary Books, 1988.

Coming of age in the milky way, Timothy Ferris, The Bodley Head, 1988.

First light: the search for the edge of the universe, Richard Preston, New American Library, 1988.

Particle physics: a Los Alamos primer, Necia Grant Cooper and Geoffrey B. West (Eds.), Cambridge University Press, 1988.

Interactions: a journey through the mind of a particle physicist and master of this world, Sheldon L. Glashow with Ben Bova, Warner Books, 1988.

Invisible matter and the fate of the universe, Barry Parker, Plenum Press, 1989.

Cosmic coincidences: dark matter, mankind and anthropic cosmology, John Gribbin and Martin Rees, Bantam, 1989.

Deep time: the journey of a single subatomic particle from the moment of creation to the death of the universe, and beyond, David Darling, Bantam, 1989.

Particle physics in the cosmos, Richard A. Carrigan and W. Peter Trower (Eds.), Scientific American and W.H. Freeman, 1989.

Physics through the 1990s: elementary-particle physics, National Academy Press, 1990.

End in fire: the supernova in the Large Magellanic Cloud, Paul Muirdin, Cambridge University Press, 1990.

Supernova: the violent death of a star, Donald Goldsmith, Oxford University Press, 1990.

Femtophysics: a short course on particle physics, M.G. Bowler, Pergamon, 1990.

Particles and forces: at the heart of the matter, Richard A. Carrigan and W. Peter Trower (Eds.), Scientific American and W.H. Freeman, 1990.

The matter myth: towards 21st-century science, P.C.W. Davies and John Gribbin, Viking, 1991.

Theories of everything: the quest for ultimate explanation, John D. Barrow, Vintage, 1991.

Lonely hearts of the cosmos: the story of the scientific quest for the secret of the universe, Dennis Overbye, Macmillan, 1991.

Atom: journey across the subatomic cosmos, Isaac Asimov, Mandarin, 1992.

Masters of time: how wormholes, snakewood and assaults on the Big Bang have brought mystery back to the cosmos, John Boslough, J.M Dent, 1992.

Stephen Hawkins: quest for a theory of everything, Kitty Ferguson, Bantam, 1992.

Genius: Richard Feynman and modern physics, James Gleick, Abacus, 1992.

The God particle: if the universe is the answer, what is the question?, Leon M. Lederman and Dick Teresi, Houghton Mifflin, 1993.

Wrinkles in time: the imprint of creation, George Smoot and Keay Davidson, Abacus, 1993.

Afterglow of creation: from the fireball to the discovery of cosmic ripples, Marcus Chown, Arrow, 1993.

In the beginning: after COBE and before the Big Bang, John Gribbin, Little, Brown & Co., 1993.

Dreams of a final theory: the search for the fundamental laws of nature, Steven Weinberg, Vintage, 1993.

Ripples of the cosmos: a view behind the scenes of the new cosmology, Michael Rowan-Robinson, W.H. Freeman, 1993.

Home is where the wind blows: chapters from a cosmologist's life, Fred Hoyle, Oxford University Press, 1994.

The hidden universe, Roger J. Tayler, Wiley–Praxis, 1995.

Exploring the X-ray universe, Philip A. Charles and Frederick D. Seward, Cambridge University Press, 1995.

Cosmology and controversy: the historical development of two theories of the universe, Helge Kragh, Princeton University Press, 1996.

Cygnus A: study of a radio galaxy, C.L. Carilli and D.E. Harris (Eds.), Cambridge University Press, 1996.

Active galactic nuclei, Ian Robson, Wiley–Praxis, 1996.

In search of the ultimate building blocks, Gerard 't Hooft, Cambridge University Press, 1997.

The whole shebang: a state-of-the-universe(s) report, Timothy Ferris, Weidenfeld & Nicolson, 1997.

The inflationary universe: the quest for a new theory of cosmic origins, Alan H. Guth, Vintage, 1998.

Large-scale structures in the universe, Anthony Fairall, Wiley–Praxis, 1998.

The astronomer's universe: stars galaxies and cosmos, Herbert Friedman, W.W. Norton, 1998.

The Victorian amateur astronomer: independent astronomical research in Britain 1820–1920, Allan Chapman, Wiley–Praxis, 1998.

Measuring the universe: the cosmological distance ladder, Stephen Webb, Springer–Praxis, 1999.

Towards the edge of the universe: a review of modern cosmology, Stuart Clark, Springer–Praxis, 1999.

In search of the Big Bang: the life and death of the universe, John Gribbin, Penguin (2nd edn.), 1999.

Measuring the universe – the historical quest to quantify space, Kitty Ferguson, Headline, 1999.

Cosmology and particle astrophysics, Lars Bergström and Ariel Goobar, Springer–Praxis, 1999.

The end of time: the next revolution in our understanding of the universe, Julian Barbour, Weidenfeld & Nicolson, 1999.

The birth of time: how we measured the age of the universe, John Gribbin, Weidenfeld & Nicolson, 1999.

Just six numbers: the deep forces that shape the universe, Martin Rees, Weidenfeld & Nicolson, 1999.

Quantum generations: a history of physics in the twentieth century, Helge Kragh, Princeton University Press, 1999.

The pleasure of finding out, Richard P. Feynman, Allen Lane/Penguin, 2000.

Cosmology revealed: living inside the cosmic egg, Anthony Fairall, Springer–Praxis, 2000.

Stardust: the cosmic recycling of stars, planets and people, John Gribbin with Mary Gribbin, Allen Lane/Penguin, 2000.

The accelerating universe: infinite expansion, the cosmological constant and the beauty of the cosmos, Mario Livio, John Wiley, 2000.

The elegant universe: superstrings, hidden dimensions and the quest for the ultimate theory, Brian Greene, Vintage, 2000.

Parallax: the race to measure the cosmos, Alan Hirshfeld, Macmillan, 2001.

The Big Bang, Joseph Silk, W.H. Freeman (3rd edn.), 2001.

The universe in a nutshell, Stephen Hawking, Bantam, 2001.

Cataclysmic variable stars: how and why they vary, Coel Hellier, Springer–Praxis, 2001.

The theory of everything: the origin and fate of the universe, Stephen Hawking, New Millennium Press, 2002.

Aeons: the search for the beginning of time, Martin Gorst, Fourth Estate, 2002.

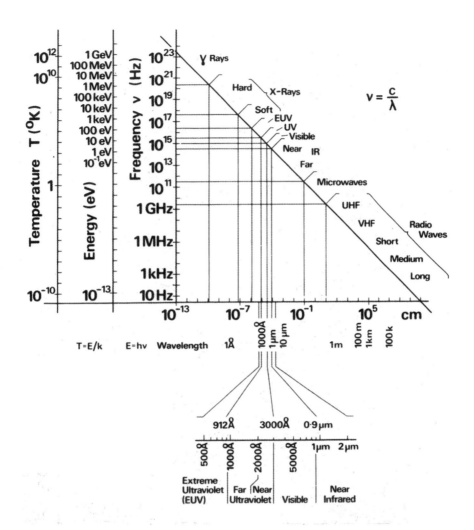

Index

Printing: Mercedes-Druck, Berlin
Binding: Stein+Lehmann, Berlin